AF251272

Medical Virology: From Pathogenesis to Disease Control

Series Editor

Shailendra K. Saxena,
Centre for Advanced Research, King George's Medical University, Lucknow, India

This book series reviews the recent advancement in the field of medical virology including molecular epidemiology, diagnostics and therapeutic strategies for various viral infections. The individual books in this series provide a comprehensive overview of infectious diseases that are caused by emerging and re-emerging viruses including their mode of infections, immunopathology, diagnosis, treatment, epidemiology, and etiology. It also discusses the clinical recommendations in the management of infectious diseases focusing on the current practices, recent advances in diagnostic approaches and therapeutic strategies. The books also discuss progress and challenges in the development of viral vaccines and discuss the application of viruses in the translational research and human healthcare.

More information about this series at http://www.springer.com/series/16573

Pranjal Chandra • Sharmili Roy

Editors

Diagnostic Strategies for COVID-19 and other Coronaviruses

 Springer

Editors
Pranjal Chandra
School of Biochemical Engineering
Indian Institute of Technology
Varanasi, India

Sharmili Roy
Division of Oncology
Stanford Medicine
California, CA, USA

ISSN 2662-981X   ISSN 2662-9828   (electronic)
Medical Virology: From Pathogenesis to Disease Control
ISBN 978-981-15-6005-7   ISBN 978-981-15-6006-4   (eBook)
https://doi.org/10.1007/978-981-15-6006-4

This Springer imprint is published by the registered company Springer Nature Singapore Pte Ltd.
The registered company address is: 152 Beach Road, #21-01/04 Gateway East, Singapore 189721, Singapore

Contents

About the Editors

Pranjal Chandra is currently serving as an assistant professor at the School of Biochemical Engineering, Indian Institute of Technology (BHU), Varanasi, India. He received his Ph.D. from Pusan National University, South Korea, and completed his postdoctoral training at Technion, Israel Institute of Technology, Israel. His research focus is highly interdisciplinary, combining biotechnology, nanobiosensors, material engineering, nanomedicine, etc. Prof. Chandra has designed several commercially viable biosensing prototypes that can be used for onsite analysis in the context of biomedical diagnostics. He has published four books on various aspects of biosensors/medical diagnostics with IET London, Springer Nature, and CRC Press USA as well as over 100 journal articles in the leading journals/books in his research area. Prof. Chandra is the recipient of many prestigious awards and fellowships, for example, the DST Ramanujan Fellowship (Government of India); ECRA (DST, Government of India); BK-21 and NRF Fellowship, South Korea; Technion Post-doctoral Fellowship, Israel; NMS Young Scientist Award; BRSI Young Scientist Award; RSC Highly Cited Corresponding Author Award (general chemistry); and ACS / Elsevier Outstanding Reviewer Awards. He is a reviewer for over 50 international journals and expert project reviewer for various national/international funding agencies. Further, Prof. Chandra is an associate editor of the journal *Sensors International* and an editorial board member of *Materials Science for Energy Technologies, World Journal of Methodology*, USA; *Frontiers of Biosciences*, USA; and *Reports in Physical Sciences*, Singapore.

Sharmili Roy is currently pursuing her second postdoctoral research in oncology/medical research in the Division of Oncology, School of Medicine, Stanford University, California, USA. She earned her Ph.D. from the University of Brunei Darussalam, Brunei, under the supervision of Prof. Minhaz Uddin Ahmed. Prior to joining Stanford University, she worked for her first postdoctoral training at the Indian Institute of Technology Guwahati, India, under supervision of Prof. Pranjal Chandra. Her current research is focused on the development of Next Generation Sequencing technology (NGS) with various clinical samples especially with many kinds of cancer DNA samples, molecular diagnostics on cancer, and DNA digital data storage process. She has expertized on point of care biosensors technology development based on loop-mediated isothermal amplification (LAMP) during her Ph.D. tenure. Dr. Roy has published several high-impact factor research papers, three USA patents, and 4 book chapters in *Biosensors and Bioelectronics*, *Analytical Methods*, *ACS Sensors*, *Food Control*, *Food Chemistry*, and *Electroanalysis*. She is a recipient of many prestigious fellowships such as: free education scholarships for master's degree studies by the Government of Sweden; Graduate Research Scholarships (GRS) from the Ministry of Education, Government of Brunei; SERB-National Post Doctorate Fellowship (NPDF) from the Department of Science and Technology (DST), Government of India; and postdoctoral fellowship from Stanford University, USA.

Insights into Novel Coronavirus and COVID-19 Outbreak

Anupriya Baranwal, Supratim Mahapatra, Buddhadev Purohit, Sharmili Roy, and Pranjal Chandra

Abstract

COVID-19, a disease caused by a virus called severe acute respiratory syndrome coronavirus 2 (SARS-COV-2), has spread across the globe, since its first outbreak in Wuhan, China. Evidence of human-to-human transmission has led to extreme quarantine measures, such as closure of borders, sealing of large cities, and confinement of people in their homes to control virus from spreading. Lessons learned from previous coronavirus outbreaks have resulted in rapid determination of virus nucleic acid sequence; however, research on SARS-COV-2 is still in nascent stage. Herein, we have compiled information on COVID-19 from recently published reports and discussed them in an organized format. We have provided brief introduction to coronavirus disease 2019 (COVID-19) and

A. Baranwal
Sir Ian Potter NanoBioSensing Facility, NanoBiotechnology Research Laboratory, School of Science, RMIT University, Melbourne, VIC, Australia

S. Mahapatra
Laboratory of Bio-Physio Sensor and Nano-Bioengineering, School of Biochemical Engineering, Indian Institute of Technology (BHU), Varanasi, UP, India

B. Purohit
Laboratory of Bio-Physio Sensor and Nano-Bioengineering, School of Biosciences and Bioengineering, Indian Institute of Technology Guwahati, Guwahati, Assam, India

S. Roy
Division of Oncology, School of Medicine, Stanford University, Stanford, CA, USA

P. Chandra (⊠)
Laboratory of Bio-Physio Sensor and Nano-Bioengineering, School of Biochemical Engineering, Indian Institute of Technology (BHU), Varanasi, UP, India

Laboratory of Bio-Physio Sensor and Nano-Bioengineering, School of Biosciences and Bioengineering, Indian Institute of Technology Guwahati, Guwahati, Assam, India

P. Chandra, S. Roy (eds.), *Diagnostic Strategies for COVID-19 and other Coronaviruses*, Medical Virology: From Pathogenesis to Disease Control, https://doi.org/10.1007/978-981-15-6006-4_1

discussed its epidemiology in detail. Moving on, we have discussed about novel coronavirus, SARS-CoV-2, and its transmission and highlighted clinical symptoms of COVID-19. In addition to this, disease diagnosis, available treatment options, and finally prevention and control measures are briefly discussed.

Keywords

Coronavirus · SARS-CoV-2 · Outbreak · COVID-19 · Infectious disease · 2019-nCoV

1   Introduction

Human being's past has been plagued with pandemics and wars; however, the dread and mortality caused by pandemics are far worse than any war. Outbreak of Spanish flu in the twentieth century was one such pandemic which infected 500 million people (ca. one-third of the world's population in 1918) and killed around 50 million people (Taubenberger et al. 2019; Taubenberger and Morens 2006). Since the beginning of twenty-first century, various virus-mediated epidemics have continuously emerged and afflicted human population across the globe. Among them, infectious diseases caused by viruses that have crossed species barrier into humans have garnered most attention. In the past two decades, outbreaks of excruciating, life-threatening respiratory diseases caused by novel coronaviruses have been observed by people worldwide (Chen 2020; Maier et al. 2015; Pan et al. 2020). Severe acute respiratory syndrome coronavirus (SARS-CoV) and Middle-East respiratory syndrome coronavirus (MERS-CoV) are two such coronaviruses that have caused >10,000 cases collectively with death rates of 10% and 37%, respectively, in the past 20 years (Luk et al. 2019; Ramadan and Shaib 2019). A comprehensive discussion on these viruses has been discussed elsewhere (Hui et al. 2016; Maier et al. 2015).

End of year 2019 saw another such outbreak when a group of patients were hospitalized in Wuhan with initial diagnosis of pneumonia from unknown aetiology (WHO 2020c). The clinical symptoms exhibited by these patients resembled with viral pneumonia. Epidemiological investigations by Wuhan Center for Disease Control and Prevention (CDC) revealed connection of these patients to a local wholesale seafood market (Wu et al. 2020b). Deep meta-transcriptomic sequencing analysis from the lower respiratory tract of the patient showed evidence of novel coronavirus which was briefly named as '2019-nCoV' by the World Health Organization (WHO) (Chang et al. 2020b). Later, International Committee on Taxonomy of Viruses announced the official name for 2019-nCoV as 'Severe Acute Respiratory Syndrome Coronavirus 2' (SARS-CoV-2) (WHO 2020b), and the disease caused by pathogen was called COVID-19. On 30th January, WHO declared COVID-19 outbreak as a "public health emergency of international concern" and later on 11th March declared it as a pandemic considering the outburst in number of confirmed cases (WHO 2020a). As of 13th May 2020, a total of 4,347,018 confirmed

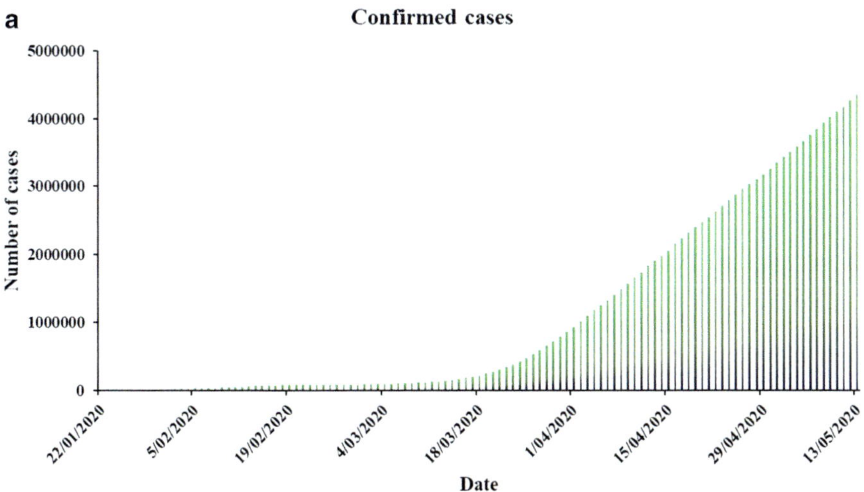

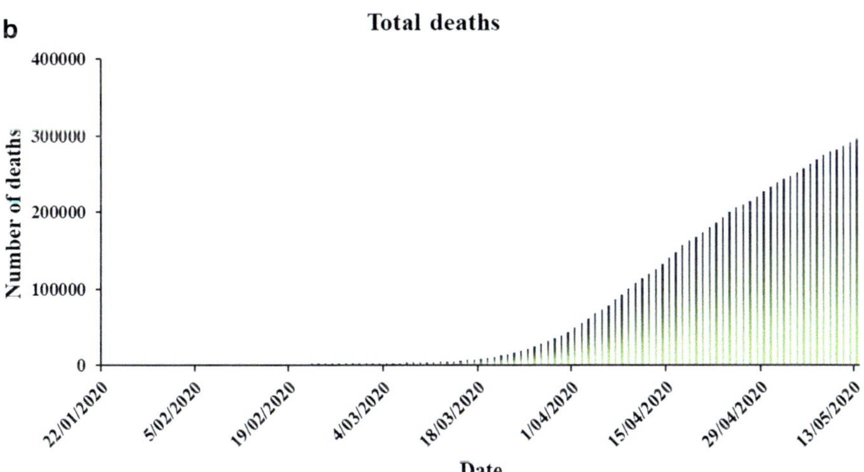

Fig. 1 Confirmed cases of COVID-19 as of 13th May 2020. Daily number of (**a**) positive cases and (**b**) total deaths across the globe. Graphs are plotted based on information published by Humanitarian Data Exchange (HDX 2020, Updated data available on: https://data.humdata.org/dataset/novel-coronavirus 2019 ncov cases)

COVID-19 cases and a total of 297,197 deaths (Fig. 1a and b) in 204 countries/ territories have been reported by the Humanitarian Data Exchange (HDX) (HDX 2020; JHU 2020). Figure 2 shows confirmed cases of COVID-19 across the globe.

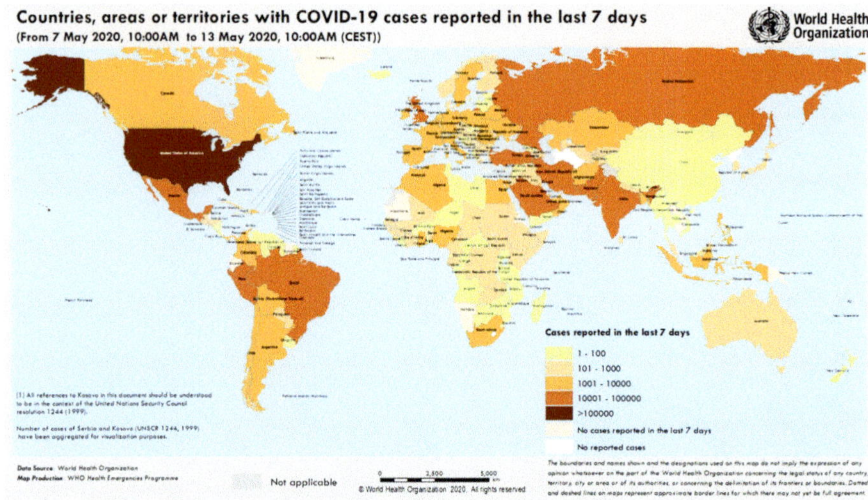

Fig. 2 Geographical distribution of reported confirmed cases of COVID-19 as of 13th May 2020. (Adapted from WHO website, ©World Health Organization 2020)

2 COVID-19 Epidemiology

According to Chinese National Health Commission (NHC), as of 24:00 on 20th February 2020, there were a total of 75,995 confirmed cases in China including 2238 deaths and 1200 cases in all five continents (outside China). The epidemiology of COVID-19 can be roughly categorized into three phases: (i) local outbreak, (ii) person-to-person transmission, and (iii) global outbreak.

(i) *Local outbreak*: The first case was reported in Wuhan, Hubei, province of China due to exposure in local wholesale seafood market. Between 18th and 29th December 2019, five patients suffering from acute respiratory distress syndrome (ARDS) were hospitalized, of which one died. The infection was confirmed to be caused by a novel coronavirus. By 2nd January 2020, the number of patients grew to 41, where <50% of these patients suffered from underlying clinical conditions like hypertension, cardiovascular disorder, and diabetes (Huang et al. 2020; Rothan and Byrareddy 2020). Epidemiological investigation revealed the supposed cause behind infection was person-to-person transmission within the hospital (nosocomial infection).

(ii) *Person-to-person transmission*: 2nd January 2020 marked commencement of the second phase due to rapid spread of infection within hospital (nosocomial infection) and between family members (close-contact transmission) (Chan et al. 2020; Li et al. 2020; Rothan and Byrareddy 2020; Xiao et al. 2020). During this phase, viral infection spreads beyond Wuhan borders and reached other cities and districts of China (Chang et al. 2020a; Liu et al. 2020). On 19th

January, new cases were reported in Beijing city and Guangdong province, and the number of confirmed cases reached 205. On 13th January 2020, the first case was reported in Thailand, transmitted by a local Wuhan resident traveling to the country. By 23rd January, 29 provinces and 6 countries reported confirmed cases marking a total of 846, which indicates ca. 20-fold increment from the first phase. In the meantime, Wuhan city prompted a lockdown to prevent movement within the city or outside. However, this period was coinciding with traditional movement of people before Chinese New Year due to which millions of people had already left Wuhan.

(iii) *Global outbreak*: 26th January 2020 marked the beginning of third phase in the form of rapid rise in cluster cases. By 30th January, the increase in number reached 240-fold, and a total of 7824 positive cases were reported. Ninety cases were reported from countries including Thailand, Taiwan, Vietnam, Sri Lanka, Malaysia, Japan, Nepal, Singapore, United Arab Emirates, the Philippines, Republic of Korea, Cambodia, the United States, India, Canada, France, Australia, Finland, and Germany. According to the news released on 30th January, 15:40 CET, the rate of mortality for COVID-19 was found to be 2.2% (170/7824) (Bassetti et al. 2020). As of 7th February 2020, 31,161 people were tested positive, and >630 people lost their lives to COVID-19 in China. By 11th February, a total of 44,730 confirmed cases were reported in China, and 441 confirmed cases were reported in 24 foreign countries. The rate of mortality in China remained high at large (1,114 deaths) when compared with the mortalities outside of China (1 death). By 16th February, this number jumped to 51,174 in China that included 1666 deaths and 15,384 severe cases. Globally, the number of confirmed cases reached to 51,857 in 25 different countries (Rothan and Byrareddy 2020). Unfortunately, as of 11th February, 1715 medical workers from 422 medical institutions were reported to have contracted the infection, of which 5 died and 1688 cases were analysed. The number of confirmed cases of SARS-CoV-2 infection has surpassed that of SARS in 2003. Among infected medical workers, majority of them were infected in Wuhan, and ca. 23% were infected in other regions of Hubei province (Surveillances 2020). The exact cause of disease transmission in medical workers and failure of personal protective equipment (PPE) seek more investigation.

Evaluation of reports from mainland China for transmission dynamics of COVID-19 showed a basic reproduction number (R0) of 2.24–3.58 (Zhao et al. 2020a). Another estimation suggested an R0 of 2.68 (Wu et al. 2020c). The estimated R_0 for SARS in 2003 in the absence of medical interventions was 2.3–3.7 (Lipsitch et al. 2003; Riley et al. 2003) which dropped to <1 by implementing rapid testing coupled with effective patient isolation (Chowell et al. 2004). This explains why SARS epidemic could be contained sooner. However, it is important to note that R_0 estimates depend on various biological, societal, and environmental factors and, therefore, must be construed with extreme caution.

3   SARS-CoV-2

Coronaviruses are large group of genetically diverse viruses that have the ability to infect a wide range of host species including mammals and birds. They belong from *Nidovirales* order and *Coronaviridae* family. The subfamily *Coronavirinae* of *Coronaviridae* family is divided into four genera: alphacoronavirus (α-CoV), betacoronavirus (β-CoV), gammacoronavirus (γ-CoV), and deltacoronavirus (δ-CoV) (Maier et al. 2015; Yang and Leibowitz 2015). Members of this subfamily can cause respiratory, hepatic, gastrointestinal, and neurological disorders. Till date, six coronaviruses capable of infecting humans have been recognized. Of these, two are from α-CoV genus (HCoVs-NL63 and HCoVs-229E), and four are from genus β-CoV (HCoVs-OC43, HCoVs-HKU1, SARS-CoV, and MERS-CoV) (Wu et al. 2020a). The novel coronavirus, SARS-CoV-2, which also belongs to β-CoV genus is the seventh member of human disease-causing coronaviruses.

SARS-CoV-2 like other coronaviruses is enveloped, non-segmented, positive-sense single-stranded RNA (+ssRNA) virus with a genome size of 30 kb (Khailany et al. 2020). It is spherical in shape with club-like surface projections that resemble solar corona (Latin corōna = crown), hence name coronavirus (Fung and Liu 2019; Maier et al. 2015). Successful isolation of SARS-CoV-2 has enabled researchers to understand its origin and pathogenicity; however, at this stage much of the information remains unknown. Viral genome sequencing has revealed that SARS-CoV-2 shares 89.1% nucleotide identity with a bat coronavirus making bats its original host (Perlman 2020; Wu et al. 2020b). In addition to this, some reports suggest snakes, pangolins, and minks to be the plausible hosts; however, subsequent studies confirmed minks and pangolins as intermediate hosts (Ji et al. 2020; Lam et al. 2020; Xu et al. 2020).

4   Transmission

Owing to the fact that first case of COVID-19 was associated with wholesale seafood market in Wuhan, animal-to-human transmission is suspected to be the actual route of transmission. However, subsequent cases that initially appeared as five patients in a family cluster did not follow the same route but rather human-to-human transmission (Phelan et al. 2020). So far, symptomatic people are believed to be the main source of transmission as the possibility of spreading infection before symptoms become apparent is quite rare, albeit not unlikely (Rothe et al. 2020; Shen et al. 2020). Therefore, asymptomatic cases should also be given importance in mediating disease transmission.

As with any respiratory pathogen, transmission of SARS-CoV-2 is mainly due to respiratory droplets emanating from sneeze and cough. In addition to this, aerosol transmission is also quite likely to aid virus spread, especially in closed spaces. Evidence of virus in stool samples of COVID-19 patients indicates that virus can survive and replicate in gastrointestinal tract. This suggests the likelihood of fecal-oral route transmission (Holshue et al. 2020); however, to confirm whether eating virus-contaminated food can cause infection, more investigations are required. Furthermore, there have been reports in media of a 30-h newborn who was tested positive for COVID-19, suggesting likelihood of in utero mother-to-fetus transmission (D'Amore 2020). However, these reports lacked information on the possibility of perinatal or postnatal modes of transmission. Recent studies published by Chen et al. (2020a) and Zhu et al. (2020a) on pregnant women and neonates report no evidence of vertical infection transmission. Therefore, at the moment, it is unknown whether there could be mother-to-fetus transmission of SARS-CoV-2. Given dearth of information, it is likely to assume that a neonate born to a COVID-19-positive mother could be infected, in utero or perinatally; however, this needs to be validated by more scientific evidence (Zhu et al. 2020a).

5   Clinical Symptoms

The clinical spectrum of SARS-CoV-2 infection varies from symptomatic or asymptomatic to severe clinical conditions like respiratory failure, septic shock, multiple organ dysfunction syndrome, cardiovascular complications, etc. Based on study published by Li et al., the clinical symptoms of COVID-19 start becoming apparent after incubation of ca. 5.2 days (Li et al. 2020). The duration from onset of symptoms to death of a patient is reported to range from 6 to 41 days with an average of 14 days. However, given this duration depends on patients' age and their immune system status, it is shorter in patients aged over 65 years (Wang et al. 2020d). In addition to this, people with underlying conditions, such as asthma, diabetes, HIV, hypertension, cancer, and heart disease, and those with compromised immune system are more predisposed to getting severely affected.

Depending on the severity of COVID-19 disease, clinical symptoms can be categorized into mild, moderate, severe, and critical forms:

(i) *Mild disease*: until now, most patients have suffered from mild form of COVID-19 and exhibited non-pneumonia or mild pneumonia symptoms. The most common symptoms displayed by patients include dry cough, sore throat, mild fever (<37.8 °C), nasal congestion, fatigue, myalgia, or headache (Cascella et al. 2020; Huang et al. 2020).

(ii) *Moderate disease*: symptoms for moderate form of disease include cough, fever (>37.8 °C), diarrhoea, nausea, vomiting, and breathlessness or tachypnoea in children without symptoms of severe pneumonia (Cascella et al. 2020; Li et al. 2020).

(iii) *Severe disease:* 14% of the total reported cases suffered from severe COVID-19 and displayed symptoms, like severe dyspnoea (noticeable shortness of breath) with respiration rate ≥ 30/min, chest discomfort, abdominal pain, acute renal damage, haemoptysis, confusion, and lymphopenia (Chen et al. 2020b; Huang et al 2020; Ren et al. 2020).

(iv) *Critical disease*: only 5% of the confirmed cases suffered from disease in critical form. They displayed symptoms like respiratory failure, acute cardiac

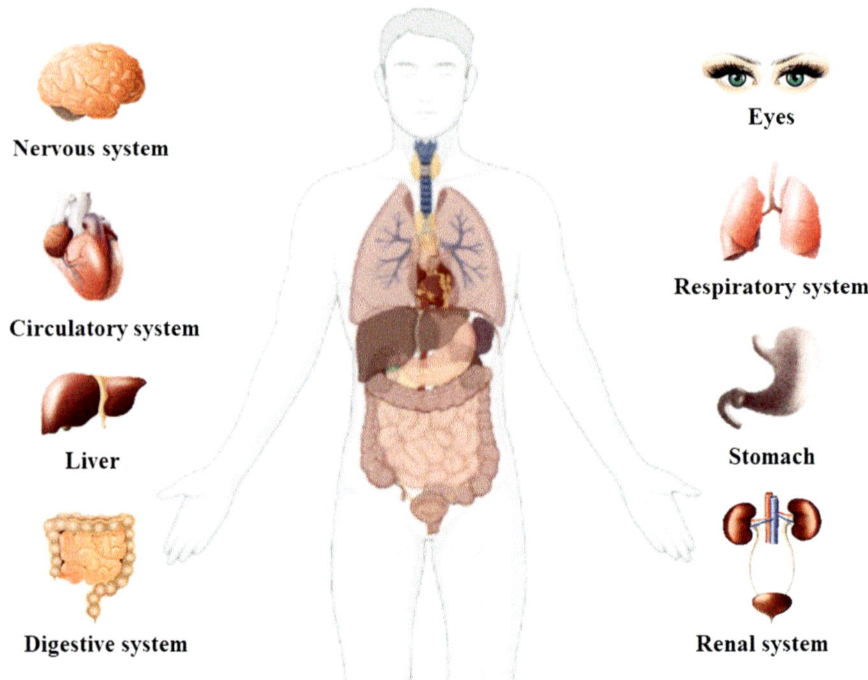

Fig. 3 Different body organs affected in a COVID-19 patient confirmed based on clinical features or biopsy of the patient

injury, acute respiratory distress syndrome, septic shock, and multiple organ dysfunction or failure (Cascella et al. 2020; Wang et al. 2020a).

It is crucial to indicate that SARS-CoV-2 infection shares similarity with earlier reported β-CoVs, SARS-CoV, and MERS-CoV, in terms of common upper respiratory tract infection (URTI) symptoms, for instance, fever, dry cough, dyspnoea, bilateral ground-glass opacities on chest CT scans, fatigue, and myalgia (Lee et al. 2003; Wang et al. 2020a; Zhu et al. 2020b).

However, serious symptoms of URTI, such as rhinorrhoea, sneezing, or sore throat, in COVID-19 patients are less common which suggests SARS-CoV-2 has higher preference for lower respiratory tract infection (Huang et al. 2020). Moreover, the chest radiographs of some patients display an infiltrate in their lungs which is related to increasing dyspnoea with hypoxemia (Phan et al. 2020). Another interesting finding suggests that gastrointestinal (GI) symptoms, particularly diarrhoea, are exhibited by COVID-19 patients. Therefore, examination of urine and faecal samples is extremely crucial to eliminate other potential routes of transmission (Holshue et al. 2020). Figure 3 shows different organs that are affected in patients infected with SARS-CoV-2 virus (Wadman et al. 2020; Wu et al. 2020d; Xiao et al. 2020).

6 Diagnosis

COVID-19 patients display nonspecific clinical symptoms due to which they cannot be used for accurate disease diagnosis. According to Guan et al., most patients in China had fever either prior to or after their hospitalization (Guan et al. 2020). Many of these patients also had cough (68%), sputum production (34%), fatigue (38%), and shortness of breath (19%) (Guan et al. 2020; Kobayashi et al. 2020; WHO 2020d). Considering the fact that many of these symptoms can be linked with other respiratory disorders, nucleic acid tests and chest imaging have been used for accurate diagnosis and verification of COVID-19. Meanwhile, serological tests are also being developed (Pang et al. 2020).

6.1 Standard Tests

Studies suggest that COVID-19 patients suffer from eosinopenia, lymphopenia, prolonged prothrombin time (PT), thrombocytopenia, and elevated levels of enzymes like alanine aminotransferase (ALT), lactate dehydrogenase (LDH), and aspartate aminotransferase (AST). In addition to this, there are elevated level of D-dimer, neutrophils, HS-troponin, and C-reactive protein (CRP) have also been reported (Bai et al. 2020; Guan et al. 2020; Huang et al. 2020; Ruan et al. 2020; Wang et al. 2020a; Zhao et al. 2020b). The most common finding from various laboratories suggests that of all aforementioned issues, eosinopenia (78.8%) and lymphopenia (68.9%) have the highest frequency in patients. It is interesting to note that although eosinopenia is associated with COVID-19 disease, its sensitivity (82%) and specificity (64%) are low. This equates to less positive and negative likelihood ratios of 2.29 and 0.28, respectively. A combination of eosinopenia and lymphopenia changes the sensitivity (38.5%) and specificity (75.5%) and worsens the positive (1.57) and negative (0.81) likelihood ratios (Xiuli Ding et al. 2020). The frequency of other conditions, like elevated AST, CRP, PT, LDH, D-dimer, ALT, and HS-troponin, is reported to be 63.4%, 60.7%, 58%, 47.2%, 46.4%, 21.3%, and 2.5%, respectively (Siordia Jr 2020). Alteration in HS-troponin level is indicative of cardiac tissue infiltration in COVID-19 patients (Driggin et al. 2020).

6.2 Nucleic Acid Tests

Owing to high specificity (ca. 100%) and no false positive results, reverse transcriptase–polymerase chain reaction (RT-PCR) is considered ideal technique for detecting COVID-19 (Corman et al. 2020). However, low sensitivity of 64% correlates with high positive likelihood ratio (LR+) of 64 but low negative likelihood ratio (LR−) of 0.3 calls for better alternatives (Ai et al. 2020; Fang et al. 2020; Xie et al. 2020). In this context, two consecutive RT-PCRs have been used to warrant accurate negative cases. According to Ai et al., RT-PCR is inclined to present positive to negative at 6.9 days and negative to positive at 5.1 days (Ai et al.

2020). Considering this, it is recommended to repeat RT-PCR tests after 3 days of an initial negative result. Factors that might affect sensitivity of one RT-PCR include low initial viral load, incorrect sampling, and variation of detection range from different manufacturers (Ai et al. 2020).

Albeit studies suggest use of two consecutive RT-PCRs to warrant accurate negative cases, scarcity of testing kits during pandemic is another challenge that makes COVID-19 diagnosis difficult. Therefore, studies suggest considering chest computer tomography (CT) scanning, if initial RT-PCR gives negative result. Reportedly, CT scans have high sensitivity (98%) regardless of low specificity (Fang et al. 2020). Considering this, initially, the general office of NHC China permitted positive CT scan results without RT-PCR to be diagnostic for SARS-CoV-2 infection. However, now it is removed from current list of recommendations (Committee 2020).

6.3 Chest Imaging

Imaging techniques may serve as a substitute to detect COVID-19 infection. Reportedly, CT scan has more potential compared to X-ray imaging to diagnose patients with COVID-19 (Ng et al. 2020; Yoon et al. 2020). Chest CT scans of patients with COVID-19 show both bilateral ground-glass opacification and bilateral ground-glass consolidation. However, ground glass opacification and consolidation are dominantly observed in early and late stage of disease, respectively (Chung et al. 2020; Wang et al. 2020c). Another important observation suggests that in most patients, more than two lobes are commonly affected, and infiltration is seen in all five lobes. Moreover, ground-glass consolidation is hardly present without opacification (Bernheim et al. 2020; Chung et al. 2020).

According to recently published reports, CT scan findings have relatively better specificity (80.5–25%) and sensitivity (84–98%) than that of RT-PCR (Ai et al. 2020; Fang et al. 2020). Based on the data from these two reports, a LR+ of 1.17 and LR– of 0.48 was produced. Another recent report also calculated these ratios by implementing CT scan algorithm and reported a LR+ of 4.3 and LR– of 0.2 (Wang et al. 2020c). Based on these studies, it can be proposed that a positive CT scan is highly suggestive of positive case of COVID-19. However, a negative CT scan result only suggests the likelihood of negative case with small to moderate confidence.

Considering the incessant growth in COVID-19 cases across the globe, development of better diagnostics is need of the hour (Mahapatra and Chandra 2020; Chandra 2020). According to the WHO, development of nucleic acid-based tests, antigen-antibody-based assays, and point-of-care detection (POCD) tests should be the prime focus of COVID-19 diagnostics research (WHO 2020d). POCD tests are inexpensive, handheld devices that allow rapid detection of patients in remote settings (Chandra 2016; Mahato et al. 2018; Purohit et al. 2020) and eventually aid in reducing burden on clinical laboratories (Mahmoudi et al. 2020). Furthermore, serological tests need to be developed to enhance surveillance efforts. Unlike nucleic acid-based tests, serological tests can mediate detection post recovery of the patient, because of which clinicians can keep track of both sick and recovered patients and have a better estimate of total COVID-19 cases (Udugama et al. 2020).

7   Treatment

Currently, there are no vaccines or antiviral drugs available to provide potential therapy to COVID-19 patients. Owing to this, symptomatic treatment and supportive therapies are adopted for the COVID-19 patients (Zumla et al. 2020). These include treatment of basic diseases, symptom relief, effective protection and supportive treatment of internal organs, respiratory support, active prevention, and control of complications. Extra attention is paid towards maintaining water and electrolyte balance and stabilizing internal environment of the body. Other treatments include antiviral therapy, Chinese medical treatment, immunomodulation therapy, convalescent plasma therapy, bronchoalveolar lavage, and blood purification. Figure 4 shows clinical strategies being followed for the treatment of COVID-19 patients.

Given that no specific drugs are available against SARS-CoV-2, the most effective research approach is to put 'old drug for new use.' Remdesivir, a nucleoside analogue prodrug for Ebola virus, has been reported to show good result in therapy of a patient with COVID-19 (Holshue et al. 2020; Mulangu et al. 2019). In vitro experiments also supported this finding and suggested remdesivir could control COVID-19. Meanwhile, chloroquine is reported to have immunomodulatory effect and potential to inhibit SARS-CoV-2 under in vitro conditions (Wang et al. 2020b). Remdesivir is undergoing clinical trials in numerous hospitals; however, its efficacy is currently uncertain. Other therapies tested against COVID-19 include arbidol, kaletra, lopinavir/ritonavir, neuraminidase inhibitors, and peptide EK1 (Lu 2020). The efficiency and safety of these candidate drugs need to be thoroughly studied in preclinical and clinical trials.

8   Prevention and Control

Prevention and control of an infectious disease like COVID-19 can be achieved by eliminating source of infection, blocking route of transmission, and protecting population prone to the disease. Given the fact that SARS-CoV-2 is transmitted through respiratory droplets and close contact, hand hygiene, social distancing, and PPE for medical staff and caregivers are helpful in controlling virus from spreading. A vaccine can effectively prevent susceptible population from COVID-19; however, there is no SARS-CoV-2-specific vaccine. Several research institutions and enterprises are working towards developing suitable vaccines, and so far, mRNA-1273 from Moderna, Ad5-nCoV from CanSino Biologicals, INO-4800 from Inovio, and LV-SMENP-DC- and pathogen-specific aAPC from Shenzhen Geno-Immune Medical Institute have moved to clinical development phase (Le et al. 2020). Based on this, it is likely to assume that SARS CoV 2 specific vaccine will be available in the near future.

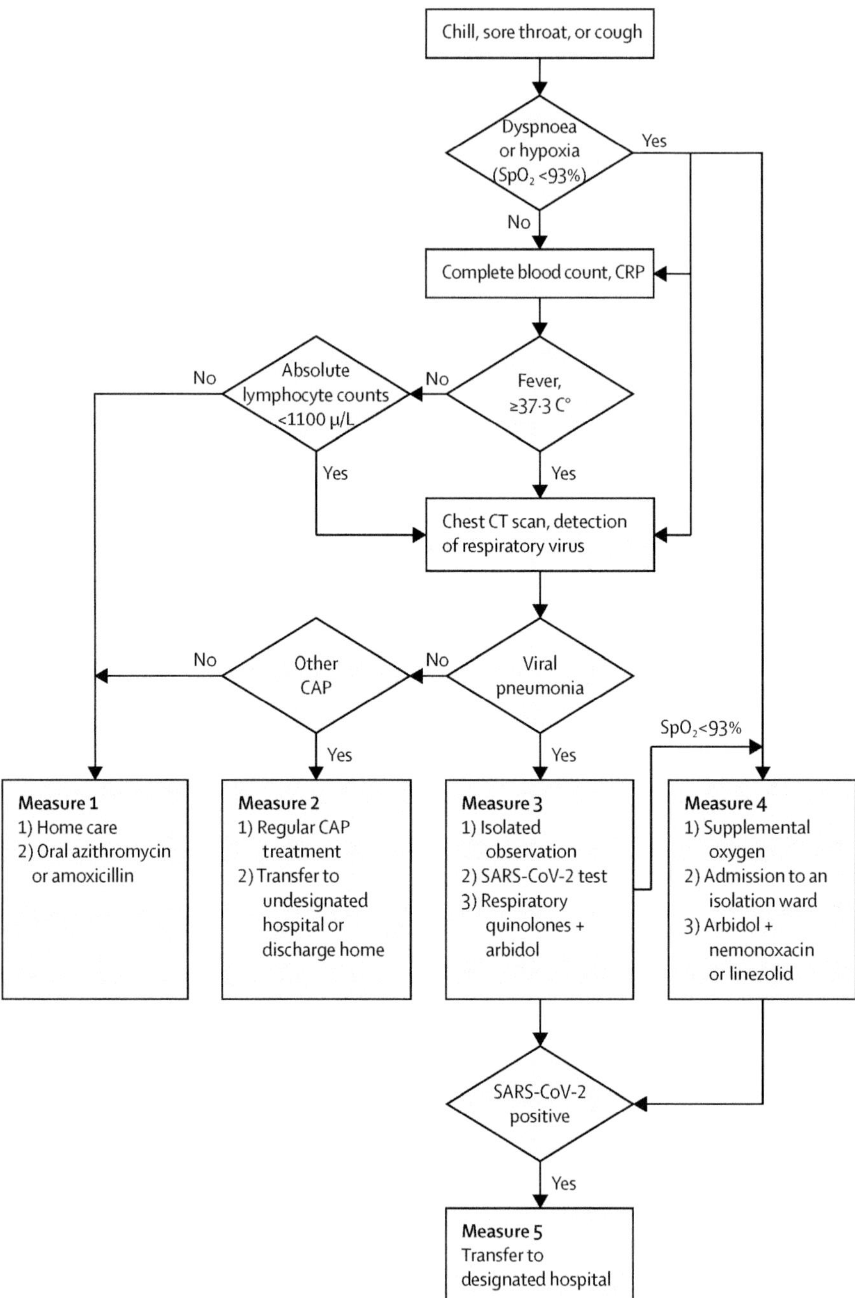

Fig. 4 Flow chart for COVID-19 patient therapy. (Figure is reproduced with permission from Zhang et al. 2020)

9 Conclusion

COVID-19 is an infectious disease caused by novel coronavirus, SARS-CoV-2. Ever since its first emergence in Wuhan, China, millions of people have tested positive across the globe, and several thousands have died. Elderly population, immunocompromised individuals, and those with underlying conditions are highly susceptible towards contracting COVID-19. Transmission of virus from one person to another can occur through respiratory droplets and close contact. The most common symptoms of COVID-19 include fever, sore throat, dry cough, fatigue, etc. Currently, our understanding of SARS-CoV-2 is rather little, except that it is highly infectious human pathogen of zoonotic origin, possibly bats. Therefore, it is crucial to understand its source of origin, route of transmission, mode of infection, interaction with host, and pathogenesis. Considering the incessant growth in number of COVID-19 cases, it is crucial to identify clinical characteristics and develop better diagnostics and effective therapeutics to overcome the ongoing pandemic. To do this, consolidated efforts from researchers, enterprises, and governments are required at local, national, and international level.

Acknowledgement Dr. Pranjal Chandra thanks Prof. Pramod Kumar Jain, Director IIT (BHU) for encouragement and providing necessary facility for completion of this work.

References

Ai T, Yang Z, Hou H, Zhan C, Chen C, Lv W, Tao Q, Sun Z, Xia L (2020) Correlation of chest CT and RT-PCR testing in coronavirus disease 2019 (COVID-19) in China: a report of 1014 cases. Radiology 200642:1–23

Bai T, Tu S, Wei Y, Xiao L, Jin Y, Zhang L, Song J, Liu W, Zhu Q, Yang L (2020) Clinical and laboratory factors predicting the prognosis of patients with COVID-19: an analysis of 127 patients in Wuhan, China. Lancet 1–17

Bassetti M, Vena A, Giacobbe DR (2020) The novel Chinese coronavirus (2019-nCoV) infections: challenges for fighting the storm. Eur J Clin Investig 50(3):e13209

Bernheim A, Mei X, Huang M, Yang Y, Fayad ZA, Zhang N, Diao K, Lin B, Zhu X, Li K (2020) Chest CT findings in coronavirus disease-19 (COVID-19): relationship to duration of infection. Radiology 295:200463

Cascella M, Rajnik M, Cuomo A, Dulebohn SC, Di Napoli R (2020) Features, evaluation and treatment coronavirus (COVID-19). Statpearls [internet] StatPearls Publishing

Chan JF W, Yuan S, Kok K H, To KK-W, Chu H, Yang J, Xing F, Liu J, Yip CC Y, Poon RW-S (2020) A familial cluster of pneumonia associated with the 2019 novel coronavirus indicating person-to-person transmission. a study of a family cluster. Lancet 395(10223):514–523

Chandra P (2016) Nanobiosensors for personalized and onsite biomedical diagnosis. The Institution of Engineering and Technology

Chandra P (2020) Miniaturised label-free smartphone assisted electrochemical sensing approach for personalised COVID-19 diagnosis. Sensors International

Chang D, Lin M, Wei L, Xie L, Zhu G, Cruz CSD, Sharma L (2020a) Epidemiologic and clinical characteristics of novel coronavirus infections involving 13 patients outside Wuhan, China. JAMA 323(11):1092–1093

Chang L, Yan Y, Wang L (2020b) Coronavirus disease 2019: coronaviruses and blood safety. Transfus Med Rev 34:75–80

Chen J (2020) Pathogenicity and transmissibility of 2019-nCoV—a quick overview and comparison with other emerging viruses. Microbes Infect 2(22):69–71

Chen H, Guo J, Wang C, Luo F, Yu X, Zhang W, Li J, Zhao D, Xu D, Gong Q (2020a) Clinical characteristics and intrauterine vertical transmission potential of COVID-19 infection in nine pregnant women: a retrospective review of medical records. Lancet 395(10226):809–815

Chen N, Zhou M, Dong X, Qu J, Gong F, Han Y, Qiu Y, Wang J, Liu Y, Wei Y (2020b) Epidemiological and clinical characteristics of 99 cases of 2019 novel coronavirus pneumonia in Wuhan, China: a descriptive study. Lancet 395(10223):507–513

Chowell G, Castillo-Chavez C, Fenimore PW, Kribs-Zaleta CM, Arriola L, Hyman JM (2004) Model parameters and outbreak control for SARS. Emerg Infect Dis 10(7):1258

Chung M, Bernheim A, Mei X, Zhang N, Huang M, Zeng X, Cui J, Xu W, Yang Y, Fayad ZA (2020) CT imaging features of 2019 novel coronavirus (2019-nCoV). Radiology 295 (1):202–207

Committee GOONH (2020) Office of state administration of traditional Chinese medicine. Notice on the issuance of a program for the diagnosis treatment of novel coronavirus infected pneumonia

Corman VM, Landt O, Kaiser M, Molenkamp R, Meijer A, Chu DK, Bleicker T, Brünink S, Schneider J, Schmidt ML (2020) Detection of 2019 novel coronavirus (2019-nCoV) by real-time RT-PCR. Eur Secur 25(3):2000045

D'Amore R (2020) Can coronavirus pass from mother to baby? Maybe, but experts need more research. Global News

Driggin E, Madhavan MV, Bikdeli B, Chuich T, Laracy J, Biondi-Zoccai G, Brown TS, Der Nigoghossian C, Zidar DA, Haythe J (2020) Cardiovascular considerations for patients, health care workers, and health systems during the COVID-19 pandemic. J Am Coll Cardiol 75 (18):2352–2371

Fang Y, Zhang H, Xie J, Lin M, Ying L, Pang P, Ji W (2020) Sensitivity of chest CT for COVID-19: comparison to RT-PCR. Radiology 296(2):200432

Fung TS, Liu DX (2019) Human coronavirus: host-pathogen interaction. Annu Rev Microbiol 73:529–557

Guan W-J, Ni Z-Y, Hu Y, Liang W-H, Ou C-Q, He J-X, Liu L, Shan H, Lei C-L, Hui DS (2020) Clinical characteristics of 2019 novel coronavirus infection in China. MedRxiv

Holshue ML, DeBolt C, Lindquist S, Lofy KH, Wiesman J, Bruce H, Spitters C, Ericson K, Wilkerson S, Tural A (2020) First case of 2019 novel coronavirus in the United States. N Engl J Med 382:929–936

Huang C, Wang Y, Li X, Ren L, Zhao J, Hu Y, Zhang L, Fan G, Xu J, Gu X (2020) Clinical features of patients infected with 2019 novel coronavirus in Wuhan, China. Lancet 395(10223):497–506

Hui DS, Rossi GA, Johnston SL (eds) (2016) SARS, MERS and other viral lung infections. European Respiratory Society Monograph, Sheffield. https://books.ersjournals.com/content/sars-mers-and-other-viral-lung-infections

Humanitarian Data Exchange (2020) Coronavirus COVID-19 cases data for China and the rest of the world. Available online: https://data.humdata.org/dataset/coronavirus-covid-19-cases-data-for-china-and-the-rest-of-the-world

Ji W, Wang W, Zhao X, Zai J, Li X (2020) Cross-species transmission of the newly identified coronavirus 2019-nCoV. J Med Virol 92:433–440. https://doi.org/10.1002/jmv.25682

John Hopkins University (2020) Coronavirus COVID-19 Global Cases by the Center for Systems Science and Engineering (CSSE) at Johns Hopkins University (JHU). Available online: https://gisanddata.maps.arcgis.com/apps/opsdashboard/index.html#/bda7594740fd40299423467b48e9ecf6

Khailany RA, Safdar M, Ozaslan M (2020) Genomic characterization of a novel SARS-CoV-2. Gene Rep 19:100682

Kobayashi T, Jung S-M, Linton NM, Kinoshita R, Hayashi K, Miyama T, Anzai A, Yang Y, Yuan B, Akhmetzhanov AR (2020) Communicating the risk of death from novel coronavirus disease (COVID-19). Multidisciplinary Digital Publishing Institute

Lam TT-Y, Shum MH-H, Zhu H-C, Tong Y-G, Ni X-B, Liao Y-S, Wei W, Cheung WY-M, Li W-J, Li L-F (2020) Identification of 2019-nCoV related coronaviruses in Malayan pangolins in southern China. BioRxiv

Le TT, Andreadakis Z, Kumar A, Roman RG, Tollefsen S, Saville M, Mayhew S (2020) The COVID-19 vaccine development landscape. Nat Rev Drug Discov 19:305–306

Lee N, Hui D, Wu A, Chan P, Cameron P, Joynt GM, Ahuja A, Yung MY, Leung C, To K (2003) A major outbreak of severe acute respiratory syndrome in Hong Kong. N Engl J Med 348 (20):1986–1994

Li Q, Guan X, Wu P, Wang X, Zhou L, Tong Y, Ren R, Leung KS, Lau EH, Wong JY (2020) Early transmission dynamics in Wuhan, China, of novel coronavirus–infected pneumonia. N Engl J Med 382:1199–1207

Lipsitch M, Cohen T, Cooper B, Robins JM, Ma S, James L, Gopalakrishna G, Chew SK, Tan CC, Samore MH (2003) Transmission dynamics and control of severe acute respiratory syndrome. Science 300(5627):1966–1970

Liu Y-C, Liao C-H, Chang C-F, Chou C-C, Lin Y-R (2020) A locally transmitted case of SARS-CoV-2 infection in Taiwan. N Engl J Med 382(11):1070–1072

Lu H (2020) Drug treatment options for the 2019-new coronavirus (2019-nCoV). Biosci Trends 14 (1):69–71

Luk HK, Li X, Fung J, Lau SK, Woo PC (2019) Molecular epidemiology, evolution and phylogeny of SARS coronavirus. Infect Genet Evol 71:21–30

Mahapatra S, Chandra P (2020) Clinically practiced and commercially viable nanobio engineered analytical methods for COVID-19 diagnosis. Biosens Bioelectron 165:112361

Mahato K, Maurya PK, Chandra P (2018) Fundamentals and commercial aspects of nanobiosensors in point-of-care clinical diagnostics. 3 Biotech 8(3):149

Mahmoudi T, de la Guardia M, Baradaran B (2020) Lateral flow assays towards point-of-care cancer detection: a review of current progress and future trends. Trends Anal Chem 125:115842

Maier HJ, Bickerton E, Britton P (2015) Coronaviruses: methods and protocols. Springer, Berlin

Mulangu S, Dodd LE, Davey RT Jr, Tshiani Mbaya O, Proschan M, Mukadi D, Lusakibanza Manzo M, Nzolo D, Tshomba Oloma A, Ibanda A (2019) A randomized, controlled trial of Ebola virus disease therapeutics. N Engl J Med 381(24):2293–2303

Ng M-Y, Lee EY, Yang J, Yang F, Li X, Wang H, Lui MM-S, Lo CS-Y, Leung B, Khong P-L (2020) Imaging profile of the COVID-19 infection: radiologic findings and literature review. Radiol Cardiothorac Imag 2(1):e200034

Pan X, Ojcius DM, Gao T, Li Z, Pan C, Pan C (2020) Lessons learned from the 2019-nCoV epidemic on prevention of future infectious diseases. Microbes Infect 22(2):86–91

Pang J, Wang MX, Ang IYH, Tan SHX, Lewis RF, Chen JI-P, Gutierrez RA, Gwee SXW, Chua PEY, Yang Q (2020) Potential rapid diagnostics, vaccine and therapeutics for 2019 novel coronavirus (2019-nCoV): a systematic review. J Clin Med 9(3):623

Perlman S (2020) Another decade, another coronavirus. Mass Med Soc 382:760–762

Phan LT, Nguyen TV, Luong QC, Nguyen TV, Nguyen HT, Le HQ, Nguyen TT, Cao TM, Pham QD (2020) Importation and human-to-human transmission of a novel coronavirus in Vietnam. N Engl J Med 382(9):872–874

Phelan AL, Katz R, Gostin LO (2020) The novel coronavirus originating in Wuhan, China: challenges for global health governance. JAMA 323(8):709–710

Purohit B, Kumar A, Mahato K, Chandra P (2020) Smartphone assisted personalized diagnostic devices and wearable sensors. Curr Opin Biomed Eng 13:42–50

Ramadan N, Shaib H (2019) Middle East respiratory syndrome coronavirus (MERS-CoV): a review. Germs 9(1):35

Ren L-L, Wang Y-M, Wu Z-Q, Xiang Z C, Guo L, Xu T, Jiang Y-Z, Xiong Y, Li Y-J, Li X-WJCMJ (2020) Identification of a novel coronavirus causing severe pneumonia in human: a descriptive study. Chin Med J 133(9):1–10

Riley S, Fraser C, Donnelly CA, Ghani AC, Abu-Raddad LJ, Hedley AJ, Leung GM, Ho L-M, Lam T-H, Thach TQ (2003) Transmission dynamics of the etiological agent of SARS in Hong Kong: impact of public health interventions. Science 300(5627):1961–1966

Rothan HA, Byrareddy SN (2020) The epidemiology and pathogenesis of coronavirus disease (COVID-19) outbreak. J Autoimmun 109:102433

Rothe C, Schunk M, Sothmann P, Bretzel G, Froeschl G, Wallrauch C, Zimmer T, Thiel V, Janke C, Guggemos W (2020) Transmission of 2019-nCoV infection from an asymptomatic contact in Germany. N Engl J Med 382(10):970–971

Ruan Q, Yang K, Wang W, Jiang L, Song J (2020) Clinical predictors of mortality due to COVID-19 based on an analysis of data of 150 patients from Wuhan, China. Intensive Care Med 46:1–3

Shen K, Yang Y, Wang T, Zhao D, Jiang Y, Jin R, Zheng Y, Xu B, Xie Z, Lin L (2020) Diagnosis, treatment, and prevention of 2019 novel coronavirus infection in children: experts' consensus statement. World J Pediatr 16:1–9

Siordia JA Jr (2020) Epidemiology and clinical features of COVID-19: a review of current literature. J Clin Virol 127:104357

Surveillances V (2020) The epidemiological characteristics of an outbreak of 2019 novel coronavirus diseases (COVID-19)—China, 2020. China CDC Wkly 2(8):113–122

Taubenberger JK, Morens DM (2006) 1918 influenza: the mother of all pandemics. Emerg Infect Dis 12(1):15

Taubenberger JK, Kash JC, Morens DM (2019) The 1918 influenza pandemic: 100 years of questions answered and unanswered. Sci Transl Med 11(502):eaau5485

Udugama B, Kadhiresan P, Kozlowski HN, Malekjahani A, Osborne M, Li VY, Chen H, Mubareka S, Gubbay J, Chan WC (2020) Diagnosing COVID-19: the disease and tools for detection. ACS Nano 14:3822–3835

Wadman M, Couzin-Frankel J, Kaiser J, Matacic C (2020) A rampage through the body. Science 368:356–360

Wang D, Hu B, Hu C, Zhu F, Liu X, Zhang J, Wang B, Xiang H, Cheng Z, Xiong Y (2020a) Clinical characteristics of 138 hospitalized patients with 2019 novel coronavirus–infected pneumonia in Wuhan, China. JAMA 323(11):1061–1069

Wang M, Cao R, Zhang L, Yang X, Liu J, Xu M, Shi Z, Hu Z, Zhong W, Xiao G (2020b) Remdesivir and chloroquine effectively inhibit the recently emerged novel coronavirus (2019-nCoV) in vitro. Cell Res 30(3):269–271

Wang S, Kang B, Ma J, Zeng X, Xiao M, Guo J, Cai M, Yang J, Li Y, Meng X (2020c) A deep learning algorithm using CT images to screen for Corona virus disease (COVID-19). MedRxiv

Wang W, Tang J, Wei F (2020d) Updated understanding of the outbreak of 2019 novel coronavirus (2019-nCoV) in Wuhan, China. J Med Virol 92(4):441–447

World Health Organization (2020a) Coronavirus disease 2019 (COVID-19): situation report, 51. Available online: https://www.who.int/docs/default-source/coronaviruse/situation-reports/20200311-sitrep-51-covid-19.pdf?sfvrsn=1ba62e57_4

World Health Organization (2020b) Naming the coronavirus disease (COVID-19) and the virus that causes it. Available online: https://www.who.int/emergencies/diseases/novel-coronavirus-2019/technical-guidance/naming-the-coronavirus-disease-(covid-2019)-and-the-virus-that-causes-it

World Health Organization (2020c) Novel Coronavirus – China. Available online: https://www.who.int/csr/don/12-january-2020-novel-coronavirus-china/en/

World Health Organization (2020d) Report of the WHO-China Joint Mission on coronavirus disease 2019 (COVID-19). 2020, 6. Available online: https://www.who.int/docs/default-source/coronaviruses/who-china-joint-mission-on-covid-19-final-report.pdf

Wu D, Wu T, Liu Q, Yang Z (2020a) The SARS-CoV-2 outbreak: what we know. Int J Infect Dis 94:44–48

Wu F, Zhao S, Yu B, Chen Y-M, Wang W, Song Z-G, Hu Y, Tao Z-W, Tian J-H, Pei Y-Y (2020b) A new coronavirus associated with human respiratory disease in China. Nature 579 (7798):265–269

Wu JT, Leung K, Leung GM (2020c) Nowcasting and forecasting the potential domestic and international spread of the 2019-nCoV outbreak originating in Wuhan, China: a modelling study. Lancet 395(10225):689–697

Wu P, Duan F, Luo C, Liu Q, Qu X, Liang L, Wu K (2020d) Characteristics of ocular findings of patients with coronavirus disease 2019 (COVID-19) in Hubei Province, China. JAMA Ophthalmol 138:575

Xiao F, Tang M, Zheng X, Liu Y, Li X, Shan H (2020) Evidence for gastrointestinal infection of SARS-CoV-2. Gastroenterology 158(6):1831–1833.e3

Xie X, Zhong Z, Zhao W, Zheng C, Wang F, Liu J (2020) Chest CT for typical 2019-nCoV pneumonia: relationship to negative RT-PCR testing. Radiology 296(2):200343

Xiuli Ding M, Geqing Xia M, Zhi Geng M, Wang Z, Wang L (2020) A simple laboratory parameter facilitates early identification of COVID-19 patients. MedRxiv

Xu X-W, Wu X-X, Jiang X-G et al (2020) Clinical findings in a group of patients infected with the 2019 novel coronavirus (SARS-Cov-2) outside of Wuhan, China: retrospective case series. BMJ:m606. https://doi.org/10.1136/bmj.m606

Yang D, Leibowitz JL (2015) The structure and functions of coronavirus genomic $3'$ and $5'$ ends. Virus Res 206:120–133

Yoon SH, Lee KH, Kim JY, Lee YK, Ko H, Kim KH, Park CM, Kim Y-H (2020) Chest radiographic and CT findings of the 2019 novel coronavirus disease (COVID-19): analysis of nine patients treated in Korea. Korean J Radiol 21(4):494–500

Zhang J, Zhou L, Yang Y, Peng W, Wang W, Chen XJTLRM (2020) Therapeutic and triage strategies for 2019 novel coronavirus disease in fever clinics. Lancet Respir Med 8(3):11–12

Zhao S, Lin Q, Ran J, Musa SS, Yang G, Wang W, Lou Y, Gao D, Yang L, He D (2020a) Preliminary estimation of the basic reproduction number of novel coronavirus (2019-nCoV) in China, from 2019 to 2020: a data-driven analysis in the early phase of the outbreak. Int J Infect Dis 92:214–217

Zhao X, Zhang B, Li P, Ma C, Gu J, Hou P, Guo Z, Wu H, Bai Y (2020b) Incidence, clinical characteristics and prognostic factor of patients with COVID-19: a systematic review and meta-analysis. MedRxiv

Zhu H, Wang L, Fang C, Peng S, Zhang L, Chang G, Xia S, Zhou W (2020a) Clinical analysis of 10 neonates born to mothers with 2019-nCoV pneumonia. Transl Pediatr 9(1):51

Zhu N, Zhang D, Wang W, Li X, Yang B, Song J, Zhao X, Huang B, Shi W, Lu R (2020b) A novel coronavirus from patients with pneumonia in China, 2019. N Engl J Med 382:727–733

Zumla A, Hui DS, Azhar EI, Memish ZA, Maeurer M (2020) Reducing mortality from 2019-nCoV: host-directed therapies should be an option. Lancet 395(10224):35–36

Advanced Biosensing Methodologies for Ultrasensitive Detection of Human Coronaviruses

Supratim Mahapatra, Anupriya Baranwal, Buddhadev Purohit, Sharmili Roy, Sanjeev Kumar Mahto, and Pranjal Chandra

Abstract

Rapid diagnosis of infectious diseases and up-to-the-minute commencement of relevant treatments are important factors that not only promote positive changes in the clinical scenario but also the health of the mass at large. Surpassing the time-consuming conventional, straightforward in vitro methods for diagnosing infectious diseases, biosensors have shown their tremendous potential in the recent era. Current developments concerning biosensing technologies bring point-of-care diagnostics to the forefront. This proves to be advantageous over

S. Mahapatra
Laboratory of Bio-Physio Sensors and Nanobioengineering, School of Biochemical Engineering, Indian Institute of Technology (BHU) Varanasi, Varanasi, India

A. Baranwal
Sir Ian Potter NanoBioSensing Facility, NanoBiotechnology Research Laboratory, School of Science, RMIT University, Melbourne, VIC, Australia

B. Purohit
Laboratory of Bio-Physio Sensor and Nano-Bioengineering, Department of Bioscience and Bioengineering, Indian Institute of Technology Guwahati, Guwahati, India

S. Roy
Division of Oncology, School of Medicine, Stanford University, Stanford, CA, USA

S. K. Mahto
Tissue Engineering and Biomicrofluidics Laboratory, School of Biomedical Engineering, Indian Institute of Technology (BHU) Varanasi, Varanasi, India

P. Chandra (✉)
Laboratory of Bio-Physio Sensors and Nanobioengineering, School of Biochemical Engineering, Indian Institute of Technology (BHU) Varanasi, Varanasi, India

Laboratory of Bio-Physio Sensor and Nano-Bioengineering, Department of Bioscience and Bioengineering, Indian Institute of Technology Guwahati, Guwahati, India

© The Editor(s) (if applicable) and The Author(s), under exclusive license to Springer Nature Singapore Pte Ltd. 2020
P. Chandra, S. Roy (eds.), *Diagnostic Strategies for COVID-19 and other Coronaviruses*, Medical Virology: From Pathogenesis to Disease Control, https://doi.org/10.1007/978-981-15-6006-4_2

conventional practices that demand centralized laboratory facilities, experienced personnel, and colossal machinery. Currently, the infectious pandemic caused by the spreading of the novel coronavirus has created an unprecedented adverse effect on both the global economy and health security. The current situation of growing cases of infection despite several measures and the unavailability of testing kits to diagnose every suspected case point toward the need of urgent upgradation of the conventional diagnostic approaches to advanced, robust, and cost-effective diagnosis. Increasing demand in viral vigilance and directive regulatory steps toward the disease transmission also reveals the need for rapid as well as sensitive devices for viral diagnosis. From the last several decades, biosensors for their noteworthy sensitivity and specificity have been considered as a promising and potent tool for precise and quantifiable detection of viruses. Current developments in genetic engineering inclusive the genetic manipulation and material engineering have introduced several approaches to enhance sensitivity, selectivity, and the overall recognition efficiency of biosensors. This chapter presents an overview of the biosensing methodologies, especially focusing on various labeled and label-free techniques that have been used in the past and are being reported in the recent era for diagnosis.

Keywords

Human coronaviruses · Biosensors · SARS-CoV · MERS-CoV · SARS-CoV-2 · Point-of-care diagnosis

1   Introduction

Since the beginning of the twentieth century, outbreaks due to viruses specifically leading to respiratory diseases have caused a major setback. The spread of such viruses at a huge epidemiological scale poses as a serious threat toward the seven billion health and wealth (Donnelly et al. 2019; Knobler et al. 2004; Memish et al. 2020; Ozili and Arun 2020; Wu et al. 2020). Tracing down the outbreak history, the severe acute respiratory syndrome coronavirus (SARS-CoV) was first reported in the year 2002. Ten years later in 2012, the middle-east respiratory syndrome coronavirus (MERS-CoV) again caused a major threat (Zumla et al. 2015). Following which, the more recent outbreak of the SARS-Coronavirus 2 (SARS-CoV-2) has affirmed the fact that despite mankind's substantial efforts to improve diagnosis, treatment, and prevention strategies toward communicable diseases for the last ten decades, novel contagions remain an inevitable challenge toward global health issues (Fauci and Morens 2012; França et al. 2013). Over the period, the foremost challenge to contain such emerging contagions includes the evolution of novel infectious promoters with the rapid spreading of diseases over the human population.

Similar to several pathogenic diagnosis procedures, methods concerning SARS, MERS, and SARS-CoV-2 detection count on various laboratory-based evaluations like electron and cryo-electron microscopy (Gui et al. 2017), in vitro growth and quantification (Coleman and Frieman 2015; Hui et al. 2004; Matsuyama et al. 2020), immunological assays (Kogaki et al. 2005; Lau et al. 2004; Lee et al. 2017), and amplification of nucleic acid (Corman et al. 2020; Shen et al. 2020; Cotten et al. 2013; Liu et al. 2020) accompanied with radiological analysis (Gogna et al. 2014; Hamimi 2016; Hosseiny et al. 2020; Jardon et al. 2019; Lin et al. 2005; Nasir et al. 2020). The basic dependency upon these in vitro diagnostic procedures is also documented with several shortcomings. For instance, microscopic analysis and radio-imaging deal with inadequate sensitivity, and culturing viral strains in vitro is rather challenging (Yu et al. 2020). Moreover, enzyme-based immunological assays and approaches for nucleic-acid amplification involving polymerase chain reaction (PCR) are often time-consuming and necessitate exhaustive sample preparation steps while falling short to produce multiple detections at a time (Ben-Assa et al. 2020; Chow 2004; Kurstjens et al. 2020). PCR also requires sophistication in sample preparation, which on rare instances might pose false-negative results for tests (Li et al. 2020a; Pan et al. 2020; Xiao et al. 2020).

To date, for diagnosing several coronaviruses, the typical protocol has been followed which involves the collection of biological samples (nasopharyngeal swab, sputum, blood and in some cases tissue swabs) and delivering to a high-equipped laboratory facility for further processing which desires involvement of practiced personnel (Chandra 2020; Chow 2004; Kurstjens et al. 2020). Till the test results become available (usually in days), clinicians deliver experimental antimicrobial treatments to the patients, which further complicates the delivery of evidence-based attention in such cases.

Biosensors break the trend of such cumbersome procedures and can be labeled as one of the best examples of a simple, miniaturized analytical tool that converts the molecular recognition of an analyte of interest into a quantifiable signal through a transducer (Bhatnagar et al. 2018; Chandra 2013; Kumar et al. 2019a, b, c; Mahato et al. 2018a, b, c; Mahato and Chandra 2019). Biosensor reportedly offers inexpensive, rapid, easy-to-detect platform with high sensitivity to effectively identify pathogens for various infectious diseases and, therefore, has proven potent to deliver point-of-care detection (Chua et al. 2011; Kashish et al. 2017; Pejcic et al. 2006; Chandra et al. 2017; Sin et al. 2014; Purohit et al. 2019a, b, c, 2020a, b; Mahato et al. 2020a, b; Kumar et al. 2019a, b, c). Moreover, it also provides a low limit of detection and device portability, consumes less energy, and demands lesser reagents (Prasad et al. 2016; Purohit et al. 2019a, b, c, 2020a, b; Mahato et al. 2019, 2020a, b; Kumar et al. 2019a, b, c). Structural upgradation and enhancement using micro \nanotechnologies have significantly improved the biosensor capability for executing complex assays (molecular, genetic, etc.) for the detection of various infectious diseases (Dai and Choi 2013; Mahato et al. 2018a, b, c; Polizzi 2019; Vaddiraju et al. 2010). Biosensor-based immunoassays enhance the detection sensitivity toward pathogen-specific antigens, while multiplex recognition of host-immune response improves the inclusive specificity compared to the conventional diagnostic procedures (Chandra 2020). Additionally, system assimilation also allows varying evaluation strategies to assimilate both pathogen specific targets

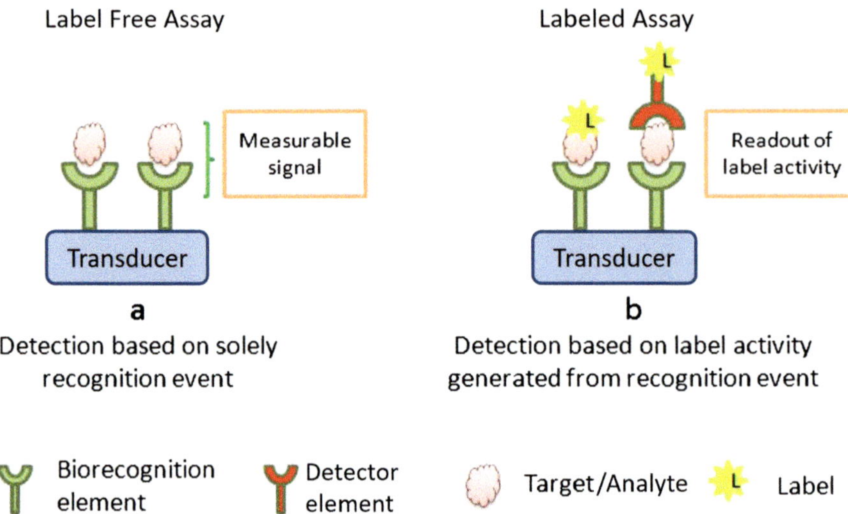

Fig. 1 Types of biosensors conventionally used for viral detection

along with biomarker's illustrative host-immune interaction responses at varying phases of infection (Mohan et al. 2011; Chandra 2016).

In this chapter, we have focused specifically on the up-to-date biosensor technologies for the detection of coronavirus diseases concentrating on various signal transducer along with their potent medical paraphrase. We have discussed the detection strategies in two distinctive sections in label-free and labeled biosensors (Fig. 1). Briefly, label-free sensors directly analyze the occurrence of an analyte via biochemical reactions (Hunt and Armani 2010; Rapp et al. 2010; Chandra 2020), whereas, for labeled sensors, the analyte is sandwiched in-between the biorecognition molecule and the explicitly labeled detector agent for signal output (Ju et al. 2011). So far, only a few reports have been published concerning biosensor-based detection of human coronaviruses. Here, we have tried to highlight those biosensors in a nutshell.

2   Label-Free Biosensor

A label-free biosensor analyzes intrinsic physicochemical property, such as charge, size, molecular weight, interfacial capacitance, resistance, refractive index, or electrical impedance of the target analyte to detect its presence in the test sample (Cooper 2009; Daniels and Pourmand 2007; Choudhary et al. 2016). Label-free biosensors warrant for a single recognition element. Assay simplification by virtue of this method is hence immense, specifically cutting reagent cost and saving assay time. Also, small molecular targets, owing to their size and reasons of obscurity within the binding region of the capturing element, are often not suitable for a labeled assay. These too can be seamlessly used with this method. Additionally, these label-free systems can carry out quantitative measurements of molecular in real-time. Considering such an advantage, continuous data recording is available. The longevity of target analytes is also increased by label-free biosensing, owing to the use of analytes in their original form devoid of any modifications. The use of label-free biosensing techniques and their applications in several infectious diseases has been discussed below, by categorizing the type of signal generated by the transducer element.

2.1   Electrical Transducer

Electrical transducer-based detection methodologies comprehensively support the broad-spectrum approaches for subtle, miniaturized, and portable biomarker recognition procedures. This is done as it can be easily integrated within standard electronic microfabrication setups and will have the swift emerging capability in microfluidics. It can also perform the simplistic yet diversified heterogeneous detection of several biomolecules even in a limited volume of samples making it utterly advantageous with minimal production cost within a bench top or handheld system (Luo and Davis 2013; Baranwal and Chandra 2018). Ishikawa et al. in a study have shown that antibody mimic proteins, a type of affinity binding agents developed by in vitro selection methods, can be employed as biorecognition elements in nanobiosensors (Ishikawa et al. 2009). Here they have developed a sensor constructed on In_2O_3 nanowires and improvised it using an antibody mimic protein (AMP), a class of affinity binding agents produced by in vitro selection techniques. They have used fibronectin-based protein as an example of such AMPs. The sensor demonstrates to have a selective affinity toward the SARS biomarker nuclear (N) protein (Fig. 2) even in subnanomolar concentration. The sensitivity of the sensing device is also comparable to the contemporary immunosensing detection techniques, favorably reduces the time consumption, and excludes any involvement

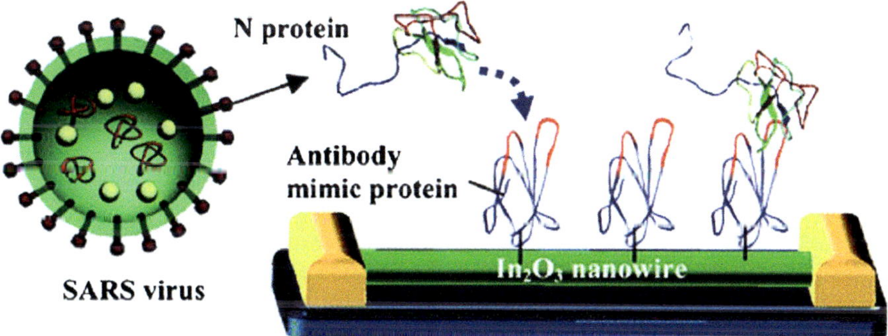

Fig. 2 Fibronectin is immobilized on the surface of an In_2O_3 nanowire as a capturing agent. The highlighted sections (red) are fibronectin with the engineered peptide sequence. It recognizes and binds the nucleocapsid (N) protein of the SARS-CoV. (Image reused with permission from Ishikawa et al. 2009)

of labeling agents. This study also demonstrated the potential of nanobiosensors. It can be used as an accurate, convenient, and rapid tool to measure the variable entities associated to complex biological systems, such as antigen-antibody, ligand-protein, oligonucleotide interaction, etc.

Again in an experiment, Layqah et al. have created and demonstrated a new one-of-a-kind competitive electrochemical immunosensor on a carbon array electrode (DEP). Further, the DEP was modified using gold nanoparticle (AuNP) which enhances the electron transfer efficiency and the electrode surface area which in turn increases the overall sensitivity of the sensor (Layqah and Eissa 2019). For a constant amount of antibody concentration added to the sample, the viral strain and the predetermined MERS-CoV protein compete with each other. The antigen-antibody-affinity interaction generates the signal necessitates for detection. Briefly, the response from the sensor was assessed by evaluating alteration in the highest amperage/peak current of the square wave voltammetry signal while gradually increasing the MERS-CoV antigen concentrations. The recorded results revealed a lower limit of detection value and a higher tendency of selectivity toward the other flu proteins with adequate stability. Moreover, the used DEP array electrodes are disposable, environment friendly, as well as cost-effective.

2.2 Plasmonic Transducer

Plasmonic biosensing provides swift, label-free probing of biological analytes in real-time. This method of biosensing can aid in the detection of very small-sized molecules at ultralow concentrations hence acting as excellent devices for point-of-care analysis (Mejía-Salazar and Oliveira 2018). These biosensors can be divided into two classes: one that uses thin metal films and the other that uses disjunct inorganic plasmon resonant nanostructures. There are different sensing techniques for each class, and additionally, a combination of both classes of the sensors can also be employed in certain platforms. The extensively used plasmonic biosensor uses "surface plasmon resonance" (SPR), which is a film-based sensor and purposed to characterize the interactions between biomolecules.

Qiu et al. recently have reported constructing a highly sensitive, canonical dual-functioning plasmonic biosensor that exhibits swift responsive diagnostic ability for SARS-CoV-2 detection (Qiu et al. 2020). The concept of the dual-functional plasmonic biosensor is to integrate the plasmonic photothermal (PPT) effect with localized surface plasmon resonance (LSPR) sensing platform on a two-dimensional gold nano-island (AuNI) chip. Here, two different wavelengths are projected at two different angles of incidence to stimulate the PPT effect and LSPR, which viciously improved the sensitivity, stability, and reliability of the device. The LSPR sensing unit is modulated to obtain real-time and label-free detection of the viral sequences of SARS-CoV-2, such as RdRp-COVID, ORF1ab-COVID, and E genes. Furthermore, the augmentation on the AuNI chips using the in situ PPT increases the specificity of the device to detect nucleic acid sequences by enhancing the hybridization kinetics. It is anticipated that associating the in situ PPT augmentation technique can precisely distinguish the sequence similarity, for RdRp genes from SARS-CoV and SARS-CoV-2. The LOD value of the biosensor was recorded as low to the concentration of 0.22 pM toward the selected SARS-CoV-2 sequences. Considering the pandemic circumstantial of COVID-19, the purposed dual-functional LSPR biosensor has claimed to deliver a consistent platform with ease of implementation. This new cutting-edge technology will certainly introduce a new trail for diagnosis besides the currently available conforming medical tests and the prolonged PCR analysis.

2.3   Field-Effect Transistor (FET)

The field-effect transistor (FET)-based biosensing devices have revealed quite a lot of advantageous properties compared to the other presently viable diagnostic methods. It has the capability of making exceptionally sensitive as well as rapid quantifications even for a little quantity of analytes (Janissen et al. 2017; Liu et al. 2019). These biosensors are determined as potentially worthwhile for medical diagnosis, point-of-care testing, and on-site detection applications. Seo et al. in a recent study have constructed a graphene-based FET biosensor functionalized with the SARS-CoV-2 spike antibody to detect the SARS-CoV-2 virus (Seo et al. 2020). Graphene-based biosensors can sense the variations of the surrounding on their surface while sustaining an optimal sensing environment intended for ultrasensitive and low-noise recognition. For the purpose, the SARS-CoV-2 spike antibody has been immobilized on the construct using a probe linker, 1-pyrenebutyric acid N-hydroxysuccinimide ester (PANHS). PANHS is an aromatic hydrocarbon and an indeed effective interface coupling agent. The sensor was subjected for in vitro studies on cultured SARS-CoV-2 strain and on associate clinical samples, i.e., nasopharyngeal swabs (Fig. 3). It claims to detect the specific SARS-CoV-2 antigen protein effectively recording the LOD of 1 fg/mL. Further, they have also reported testing the specificity of the sensor by distinguishing the SARS-CoV-2 antigen protein from the MERS-CoV efficiently.

3   Labeled Biosensors

Label-based biosensing essays are one of the most popular and prospering methods of biosensing nowadays. In label-based biosensors, the analyte molecule gets sandwiched between an immobilized capturing agent on a solid surface (i.e., micro/nanoparticles, electrodes, sensor-chips, etc.) and detecting agents which are typically tagged with signaling molecules, like luminescence molecules, nanoparticles, enzymes, or fluorophores (Baranwal et al. 2016; Chandra et al. 2010). Generally, the receptor (capture and detector) molecules have distinguished binding sites, which as a result enhances device specificity and reduces background noise. A typical example of ELISA-based biosensing assay uses sandwich immuno-assay for the diagnosis of various infectious diseases. It uses an antibody as a

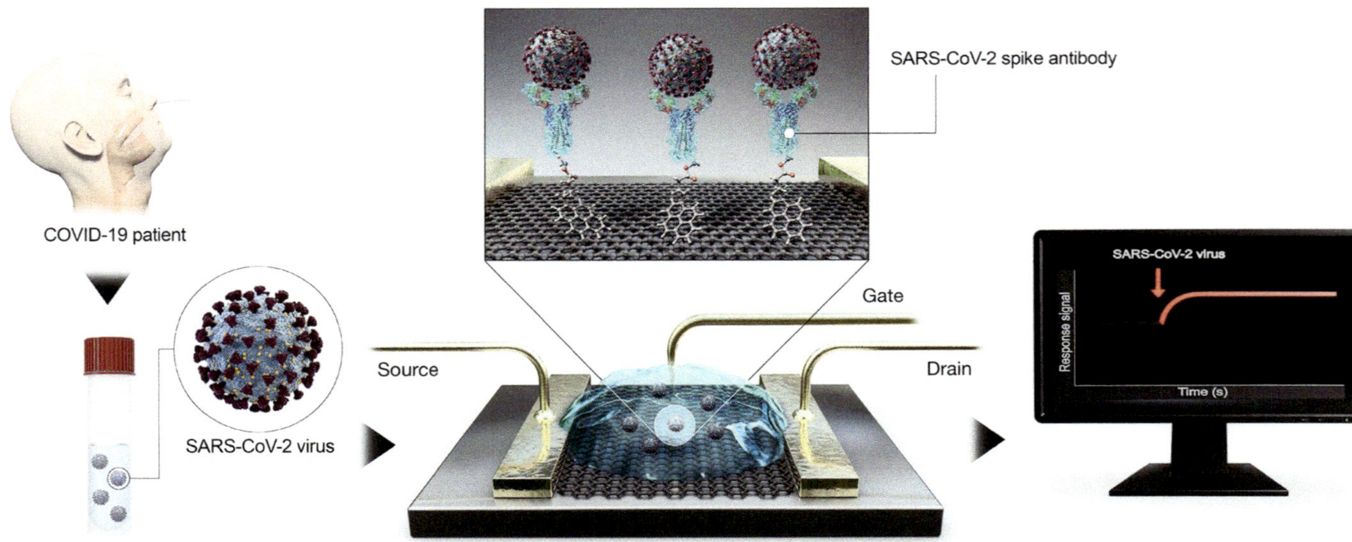

Fig. 3 Abstract illustration of a plasmonic biosensor evaluated for SARS-CoV-2 detection. (Image reused with permission from Seo et al. 2020)

captured molecule and another enzymatically modified fluorescence tagged antibody to catalyze the translation reaction of the chromogenic substrate to a visually distinctive colored product. The formed color products are then compared based on their optical densities and the concentration of analytes.

Label-based biosensors explore mechanical, optical, and electrical transducers attached to a signal-tagged molecule. The example of such interactions between sensor and tag includes optical, magneto-resistive, and electrochemical biosensors. Further, detection of fluorescence, luminescent tags, and colorimetric evaluation is done by optical sensors. Electrochemical biosensors detect redox reactions with the help of enzyme tags, whereas magnetically tag-aided detections are done by magneto-resistive biosensors. These assemblies quantitatively and semi-quantitatively detect analytes by relating the signal produced to the amount of analyte captured.

3.1   Optical Biosensor

Optical biosensors are applied to a wide array of performance in the field of detection. These not only include the detection of biological systems to promote ground-breaking advances in diagnostics but also improvements in drug discoveries and environmental supervision. Including the additional advantages of higher sensitivity, reliability, robustness, and integrity, optical biosensors also support to avoid the complication of pretreatment and probable influence on the nature of target molecules. Antibody-based immunosensors are the most viable optical sensors that are employed for the detection of pathogens (Byrne et al. 2009). Polyclonal, monoclonal, and recombinant antibodies have been often selected for immunodiagnostics and biomarker detection. This technique is known as an immunochromatographic test (ICT) (Kogaki et al. 2005). ICT works generally based on a sandwich format by means of double antigens or double antibodies. For the rapid detection of SARS-CoV, Tyson Bioresearch, Inc., Taipei, Taiwan, developed an immunogold-based ICT device by incorporating the recombinant N protein antigen of SARS-CoV in the test (Wu et al. 2004). The construction of the ICT device is such that a nitrocellulose strip is present, wherein a detection zone is allocated at the top. This is where an anti-mouse IgG and SARS-CoV N protein is present in an immobilized state onto the control and test line. In the middle, the strip consists of mouse IgG and SARS CoV N protein teamed with gold nanoparticles serving as the detector/locator. Two wells for the sample and the buffer are present at the bottom. For the assay, a neat serum sample and testing buffer are added to the sample and the buffer well, respectively. An antibody-antigen-gold complex is formed if the sample comprises specific antibodies to SARS-CoV. After lateral flow along the membrane, the colored complex of antibodies antigen gold gets accumulated on the test line, and a red color becomes apparent to the naked eye. Two parallel red lines are seen on a positive result; the control line implies that the device is working fine, and the test line indicates the presence of the SARS-CoV antibody in the serum sample. In case of a contrary result, the red line will only be seen on the control region. The test is invalid if red color is found only at the test line or no lines are visible at all. Following a very similar mechanism, Sino Biological, Beijing, China, had released the first ICT kits for the detection of SARS-CoV-2 N and S proteins (Web reference 1).

Huang et al. previously in a study explained the generation of a dual-monoclonal-antibody system tagged with glutathione S-transferase (GST) to apply against SARS-CoV N protein (Huang et al. 2004). Detection of a low concentration GST-N protein (15 ng/mL and 1 ng/mL in PBS and diluted serum, respectively) was accomplished by the use of conventional antigen capture ELISA initially. Later on, an ultrasensitive localized surface plasmon coupled fluorescence (LSPCF) fiberoptic biosensor was developed to further augment the process of detection. After this development, the detection of GST-N in PBS was recorded at the lowest concentration of 0.1 pg/mL. The LOD recorded for 10-fold diluted human serum was 0.1 pg/mL, that is, comparable to that of in raw serum sample (1 pg/mL). The application of LSPCF enhanced the sensitivity of detection by 10^4-fold by making use of the same monoclonal antibodies. Taking into account the limit of detection and cost-effectiveness, the use of LSPCF always proved to be a preferred method for SARS-CoV N protein detection from serum sample. Fluorescence can be enhanced and excited by LSPs with greater efficiency close to the GNP surface. As over 40 fluorophores excite simultaneously that are presented on each fluorescence probe collectively amplifies the fluorescence signal. In the coming era, LSPCF will also show immense possibilities through utilizing the promising chip-based evaluation to measure serum protein levels both qualitatively and quantitatively. On contrasting with conventional setups, the LSPCF fiber-optic biosensors can detect fluorescence signals near the reaction region and hence accentuate the collection of fluorescence (Chang et al. 2009; Hsieh et al. 2007). Owing to this elevated quantitative capability and sensitivity, this technique can be employed for easy and prompt identification of diseases.

Recently, based on lateral flow assay, Li et al. developed a point-of-care immunoassay kit to detect IgM and IgG antibodies simultaneously in human blood within 15 min (Fig. 4) (Z. Li et al. 2020b). They had reported a clinical trial involving eight hospitals and Chinese CDC agencies corroborating the kit's medical efficiency. The results were promising and show the rapid detection of antibodies with greater sensitivity and specificity. Owing to excellent results, it is now being authorized to be used in hospitals, clinics, and laboratories and, thus, become a compelling medium in the fight against SARS-CoV-2.

3.2 Genosensor

In the recent era, the generations of nano-biosensors and their different usage protocol have shown a curve of evolution nevertheless. Such an example of a sensor is a genosensor. A genosensor is a typical gene-based sensor that exploits immobilized genetic probes as the recognition element to evaluate specific binding processes that involves the formation of hybrids, i.e., DNA–DNA or DNA–RNA

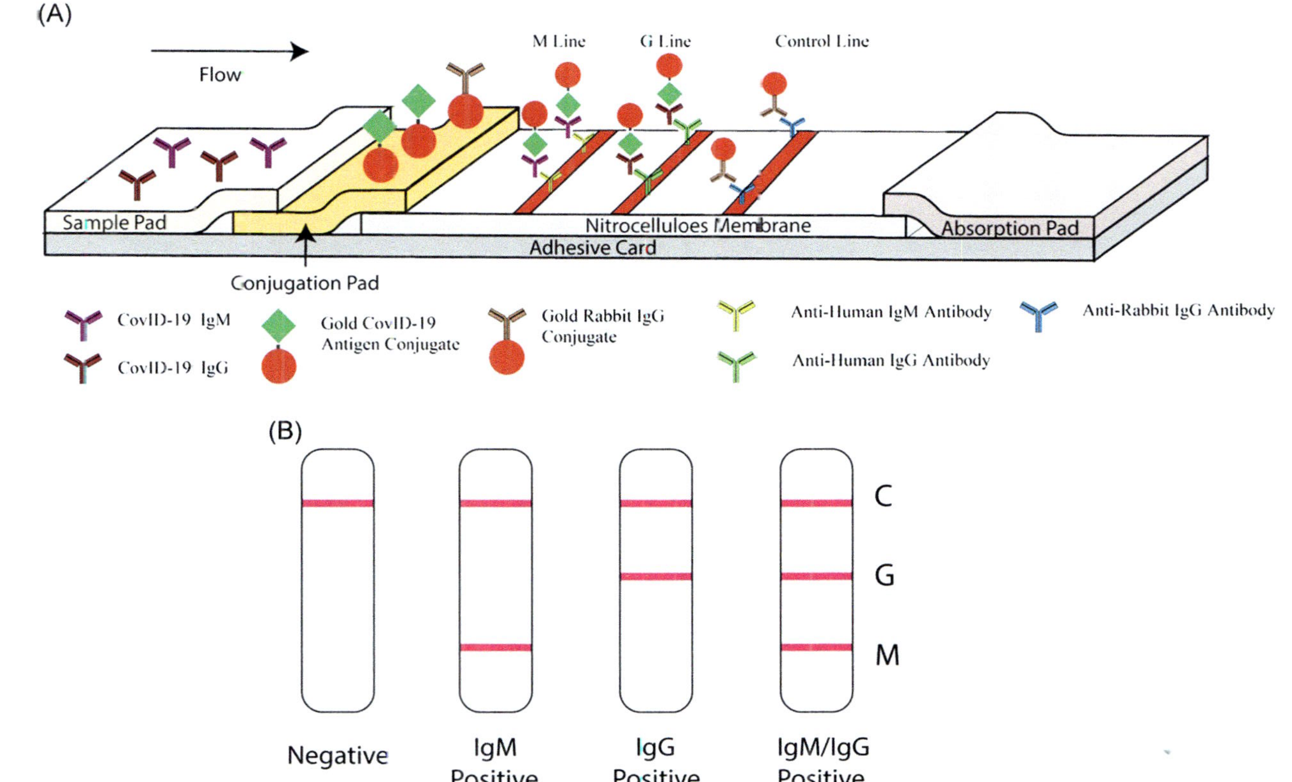

Fig. 4 Graphical representation of rapid IgM-IgG-combined antibody test for SARS-CoV-2. A typical detection device (**a**); abstract representation of varying testing results (**b**). G stands for immunoglobulin G (IgG); M stands for immunoglobulin M (IgM); C is the control line. (Image reused with permission from Li et al. 2020a, b)

hybrids, and the interactions between ligand molecules or proteins and DNA at the sensing surface. Further, genosensors can be classified into the following types. A receptor-based genosensor, consisting of a bioreceptor probe, that is typically small oligonucleotide sequences. In the case of an aptamer-based genosensor, the aptamer is made up of a sequence of synthetic oligonucleotides that is rendered immobile at the transducer surface, and in due course of time for their affinity, these oligonucleotide sequences acknowledge the nucleic acids analyte by creating complementary duplexes. Additionally, nanocomposites/nanoparticles can be tagged and coupled with genosensors to collectively improve the process of immobilization and the overall sensitivity of the oligonucleotide sequence on the surface of the transducer. In a study, Abad-Valle et al. had reported designing a simple, sensitive, and cost-effective miniaturized homemade device. It was based on a structural framework made of thin gold film for electrochemical detection in fewer volumes (Abad-Valle et al. 2005). They employed the construct for further development of genosensors especially in the detection of the specific sequence of a SARS-CoV virus. The DNA probe was immobilized on the gold surface, and the process was carried out through a thiol group linked at the $3'$ end with an aliphatic spacer. The parameters that affected the immobilization were studied using a double-labeled (biotinylated and thiolated) DNA strand. Enzymatic detection was carried out by alkaline phosphatase-labeled streptavidin. Also, blocking with 1-hexanethiol produced well-defined signals. Subsequently, the solvent was evaporated to achieve optimum results, favoring immobilization while avoiding hybridization. A low limit of detection (6 pM) concerning the previously reported analogous schemes in literature was achieved on enzymatic hydrolysis of 3-indoxyl phosphate. The specificity of the assessment was verified using a 3-base mismatch DNA strand, where a strong inequity was reported while retaining 1 h of hybridization period and 50% of formamide in the buffer. The group has also specified to study in detail the processes to discriminate the 2-base and 1-base mismatching strands as a prospect.

3.3 Luciferase-Based Sensors

Kilianski et al. have reported developing a luciferase-based biosensor to detect MERS-CoV. They employed the biosensor construct for expressing the two MERS-CoV-specific biomarkers, the papain-like protease (PLpro), and the 3-chymotrypsin-like protease (3CLpro) while monitoring their activity in vitro simultaneously (Kilianski et al. 2013). It has been demonstrated that the biosensor recognizes the expressed PLpro in MERS-CoV while processing the recognized CoV-PLpro cleavage site, RLKGG. However, because of the divergent amino acid sequence in the binding site of the drug, the already in-use CoV-PLpro inhibitors were not able to block MERS-CoV PLpro activity. Again, they utilize the luciferase-based biosensors together with the recognized 3CLpro cleavage site VRLQS to understand the activity of 3CLpro, by expressing the protease affixed with nonstructural protein 4 (nsp4) and the amino-terminal portion of nsp6. They have also determined that similar to SARS-CoV and murine CoV, a small-molecule inhibitor inhibits the replication process in the case of MERS-CoV 3CLpro and inhibits the activity. As a whole, the developed biosensor assays involving the proteases permit rapid identification as well as the evaluation of the protease inhibitors and viral protease activity, respectively. It is anticipated that to assess protease activity, the expression of MERS-CoV PLpro and 3CLpro is effective. Luciferase-based biosensor supports such expression of PL and CLpro that in turn enable swift identification of the replication inhibiting small molecule that is specific to MERS-CoV. Accordingly, it may also prove its compelling aspects over other coronaviruses.

4   Conclusion

Biosensing methodologies have shown enormous advancements for viral detection in terms of rapid analysis with low LOD, wide linear detection range, high sensitivity, and specificity. The need to develop point-of-care diagnostic devices for rapid, sensitive, and cost-effective screening even for multiple samples at a time explains the necessity of biosensors. It also rationalizes the increase in demand for compelling epidemiological surveillance along with high-throughput screening tests. So far biosensing techniques despite having such beneficial properties remain inceptive for commercialized diagnosis of infectious diseases. Biosensors when compared with other diagnostics come with an added advantage of being one of the most compact and handheld devices. This makes biosensors probably the best among the race of diagnostic tools. Infectious diseases in urban and suburban areas spread like wildfire in a very short duration of time; hence regulation of such outbreaks becomes a reason for concern. Hence, such rapid miniaturized tests may overcome such delinquency with effective enactment toward initial disease control and would also waive off delayed diagnosis as seen in conventional procedures. Furthermore, novel and technologically advanced nano-sized materials, composites, and polymers with significant biocompatibility also are recognized to provide improved specificity and stability to affinity reagents. It is for a fact that the current advances in technology would accelerate the development of biosensors into such a tool of profound magnificence that it will be a milestone in diagnostics and infection surveillance in medical setups and laboratories.

References

Abad-Valle P, Fernández-Abedul MT, Costa-García A (2005) Genosensor on gold films with enzymatic electrochemical detection of a SARS virus sequence. In: Biosensors and bioelectronics. https://doi.org/10.1016/j.bios.2004.10.019

Baranwal A, Chandra P (2018) Clinical implications and electrochemical biosensing of monoamine neurotransmitters in body fluids, in vitro, in vivo, and ex vivo models. Biosens Bioelectron https://doi.org/10.1016/j.bios.2018.09.002

Baranwal A, Mahato K, Srivastava A, Maurya PK, Chandra P (2016) Phytofabricated metallic nanoparticles and their clinical applications. RSC Adv. https://doi.org/10.1039/c6ra23411a

Ben-Assa N, Naddaf R, Gefen T, Capucha T, Hajjo H, Mandelbaum N, Elbaum L, Rogov P, King DA, Kaplan S, Rotem A, Chowers M, Szwarcwort-Cohen M, Paul M, Geva-Zatorsky N (2020) SARS-CoV-2 on-the-spot virus detection directly from patients. medRxiv. https://doi.org/10.1101/2020.04.22.20072389

Bhatnagar I, Mahato K, Ealla KKR, Asthana A, Chandra P (2018) Chitosan stabilized gold nanoparticle mediated self-assembled gliP nanobiosensor for diagnosis of invasive Aspergillosis. Int J Biol Macromol. https://doi.org/10.1016/j.ijbiomac.2017.12.084

Byrne B, Stack E, Gilmartin N, O'Kennedy R (2009) Antibody-based sensors: principles, problems and potential for detection of pathogens and associated toxins. Sensors (Switzerland). https://doi.org/10.3390/s90604407

Chandra P (2013) Miniaturized multiplex electrochemical biosensor in clinical bioanalysis. J Bioanal Biomed. https://doi.org/10.4172/1948-593X.1000e122

Chandra P (2016) Nanobiosensors for personalized and onsite biomedical diagnosis. Nanobiosens Pers Onsite Biomed Diagn. https://doi.org/10.1049/PBHE001E

Chandra P (2020) Miniaturized label-free smartphone assisted electrochemical sensing approach for personalized COVID-19 diagnosis. Sens Int. https://doi.org/10.1016/j.sintl.2020.100019

Chandra P, Das D, Abdelwahab AA (2010) Gold nanoparticles in molecular diagnostics and therapeutics. Dig J Nanomater Biostruct 5:363–367

Chandra P, Tan YN, Singh SP (2017) Next generation point-of-care biomedical sensors technologies for cancer diagnosis. Next Gener Point-of-care Biomed Sens Technol Cancer Diagn. https://doi.org/10.1007/978-981-10-4726-8

Chang YF, Chen RC, Lee YJ, Chao SC, Su LC, Li YC, Chou C (2009) Localized surface plasmon coupled fluorescence fiber-optic biosensor for alpha-fetoprotein detection in human serum. Biosens Bioelectron. https://doi.org/10.1016/j.bios.2008.08.019

Choudhary M, Yadav P, Singh A, Kaur S, Ramirez-Vick J, Chandra P, Arora K, Singh SP (2016) CD 59 targeted ultrasensitive electrochemical immunosensor for fast and noninvasive diagnosis of oral cancer. Electroanalysis 28:2565–2574

Chow CB (2004) Post-SARS infection control in the hospital and clinic. Paediatr Respir Rev. https://doi.org/10.1016/j.prrv.2004.07.006

Chua AL, Yean CY, Ravichandran M, Lim BH, Lalitha P (2011) A rapid DNA biosensor for the molecular diagnosis of infectious disease. Biosens Bioelectron. https://doi.org/10.1016/j.bios.2011.02.040

Coleman CM, Frieman MB (2015) Growth and quantification of MERS-CoV infection. Curr Protoc Microbiol. https://doi.org/10.1002/9780471729259.mc15e02s37

Cooper MA (2009) Label-free biosensors: techniques and applications. https://doi.org/10.1017/CBO9780511626531

Corman VM, Landt O, Kaiser M, Molenkamp R, Meijer A, Chu DKW, Bleicker T, Brünink S, Schneider J, Schmidt ML, Mulders DGJC, Haagmans BL, Van Der Veer B, Van Den Brink S, Wijsman L, Goderski G, Romette JL, Ellis J, Zambon M, Peiris M, Goossens H, Reusken C, Koopmans MPG, Drosten C (2020) Detection of 2019 novel coronavirus (2019-nCoV) by real-time RT-PCR. Eur Secur. https://doi.org/10.2807/1560-7917.ES.2020.25.3.2000045

Cotten M, Watson SJ, Kellam P, Al-Rabeeah AA, Makhdoom HQ, Assiri A, Aal-Tawfiq J, Alhakeem RF, Madani H, AlRabiah FA, Al Hajjar S, Al-Nassir WN, Albarrak A, Flemban H, Balkhy HH, Alsubaie S, Palser AL, Gall A, Bashford-Rogers R, Rambaut A, Zumla AI, Memish ZA (2013) Transmission and evolution of the Middle East respiratory syndrome coronavirus in Saudi Arabia: a descriptive genomic study. Lancet. https://doi.org/10.1016/S0140-6736(13)61887-5

Dai C, Choi S (2013) Technology and applications of microbial biosensor. Open J Appl Biosens. https://doi.org/10.4236/ojab.2013.23011

Daniels JS, Pourmand N (2007) Label-free impedance biosensors: opportunities and challenges. Electroanalysis. https://doi.org/10.1002/elan.200603855

Donnelly CA, Malik MR, Elkholy A, Cauchemez S, Van Kerkhove MD (2019) Worldwide reduction in MERS cases and deaths since 2016. Emerg Infect Dis. https://doi.org/10.3201/eid2509.190143

Fauci AS, Morens DM (2012) The perpetual challenge of infectious diseases. N Engl J Med. https://doi.org/10.1056/NEJMra1108296

França RFO, Da Silva CC, De Paula SO (2013) Recent advances in molecular medicine techniques for the diagnosis, prevention, and control of infectious diseases. Eur J Clin Microbiol Infect Dis. https://doi.org/10.1007/s10096-013-1813-0

Gogna A, Tay KH, Tan BS (2014) Severe acute respiratory syndrome: 11 years later – a radiology perspective. Am J Roentgenol. https://doi.org/10.2214/AJR.14.13062

Gui M, Liu X, Guo D, Zhang Z, Yin CC, Chen Y, Xiang Y (2017) Electron microscopy studies of the coronavirus ribonucleoprotein complex. Protein Cell. https://doi.org/10.1007/s13238-016-0352-8

Hamimi A (2016) MERS-CoV: Middle East respiratory syndrome corona virus: can radiology be of help? Initial single center experience. Egypt J Radiol Nucl Med. https://doi.org/10.1016/j.ejrnm.2015.11.004

Hosseiny M, Kooraki S, Gholamrezanezhad A, Reddy S, Myers L (2020) Radiology perspective of coronavirus disease 2019 (COVID-19): lessons from severe acute respiratory syndrome and Middle East respiratory syndrome. AJR Am J Roentgenol. https://doi.org/10.2214/AJR.20.22969

Hsieh BY, Chang YF, Ng MY, Liu WC, Lin CH, Wu HT, Chou C (2007) Localized surface plasmon coupled fluorescence fiber-optic biosensor with gold nanoparticles. Anal Chem. https://doi.org/10.1021/ac0624389

Huang Q, Yu L, Petros AM, Gunasekera A, Liu Z, Xu N, Hajduk P, Mack J, Fesik SW, Olejniczak ET (2004) Structure of the N terminal RNA-binding domain of the SARS CoV nucleocapsid protein. Biochemistry. https://doi.org/10.1021/bi036155b

Hui DSC, Chan MCH, Wu AK, Ng PC (2004) Severe acute respiratory syndrome (SARS): epidemiology and clinical features. Postgrad Med J 80:373 LP–373381. https://doi.org/10.1136/pgmj.2004.020263

Hunt HK, Armani AM (2010) Label-free biological and chemical sensors. Nanoscale. https://doi.org/10.1039/c0nr00201a

Ishikawa FN, Chang H, Curreli M, Liao H, Olson KCA, Chen P, Zhang R, Roberts RW, Sun R, Cote KRJ, Thompson ME, Zhou C (2009) Label-free, electrical detection of the SARS virus N-protein with nanowire capture probes. ACS Nano

Janissen R, Sahoo PK, Santos CA, Da Silva AM, Von Zuben AAG, Souto DEP, Costa ADT, Celedon P, Zanchin NIT, Almeida DB, Oliveira DS, Kubota LT, Cesar CL, Souza APD, Cotta MA (2017) InP nanowire biosensor with tailored Biofunctionalization: ultrasensitive and highly selective disease biomarker detection. Nano Lett. https://doi.org/10.1021/acs.nanolett.7b01803

Jardon M, Mohammad SF, Jude CM, Pahwa A (2019) Imaging of emerging infectious diseases. Curr Radiol Rep. https://doi.org/10.1007/s40134-019-0338-4

Ju H, Zhang X, Wang J (2011) Signal amplification for Nanobiosensing. https://doi.org/10.1007/978-1-4419-9622-0_2

Kashish BS, Jyoti A, Mahato K, Chandra P, Prakash R (2017) Highly sensitive in vitro biosensor for Enterotoxigenic Escherichia coli detection based on ssDNA anchored on PtNPs-chitosan Nanocomposite. Electroanalysis. https://doi.org/10.1002/elan.201600169

Kilianski A, Mielech AM, Deng X, Baker SC (2013) Assessing activity and inhibition of Middle East respiratory syndrome coronavirus papain-like and 3C-like proteases using luciferase-based biosensors. J Virol. https://doi.org/10.1128/jvi.02105-13

Knobler S, Mahmoud A, Lemon S, Mack A, Sivitz L, Oberholtzer K (2004) Learning from SARS: preparing for the next disease outbreak. The National Academies Press, Washington

Kogaki H, Uchida Y, Fujii N, Kurano Y, Miyake K, Kido Y, Kariwa H, Takashima I, Tamashiro II, Ling AE, Okada M (2005) Novel rapid immunochromatographic test based on an enzyme immunoassay for detecting nucleocapsid antigen in SARS-associated coronavirus. J Clin Lab Anal. https://doi.org/10.1002/jcla.20070

Kumar A, Purohit B, Maurya PK, Pandey LM, Chandra P (2019a) Engineered nanomaterial assisted signal-amplification strategies for enhancing analytical performance of electrochemical biosensors. Electroanalysis. https://doi.org/10.1002/elan.201900216

Kumar A, Purohit B, Mahato K, Mandal R, Srivastava A, Chandra P (2019b) Gold-iron bimetallic nanoparticles impregnated reduced Graphene oxide based Nanosensor for label-free detection of biomarker related to non-alcoholic fatty liver disease. Electroanalysis. https://doi.org/10.1002/elan.201900337

Kumar A, Roy S, Srivastava A, Naikwade MM, Purohit B, Mahato K, Naidu VGM, Chandra P (2019c) Chapter 10 – Nanotherapeutics: a novel and powerful approach in modern healthcare system. Nanotechnol Mod Anim Biotechnol. https://doi.org/10.1016/B978-0-12-818823-1.00010-7

Kurstjens S, van der Horst A, Herpers R, Geerits MWL, Kluiters-de Hingh YCM, Göttgens E-L, Blaauw MJT, Thelen MHM, Elisen MGLM, Kusters R (2020) Rapid identification of SARS-CoV-2-infected patients at the emergency department using routine testing. https://doi.org/10.1101/2020.04.20.20067512

Lau SKP, Woo PCY, Wong BHL, Tsoi HW, Woo GKS, Poon RWS, Chan KH, Wei WI, Malik Peiris JS, Yuen KY (2004) Detection of severe acute respiratory syndrome (SARS) coronavirus nucleocapsid protein in SARS patients by enzyme-linked immunosorbent assay. J Clin Microbiol. https://doi.org/10.1128/JCM.42.7.2884-2889.2004

Layqah LA, Eissa S (2019) An electrochemical immunosensor for the corona virus associated with the Middle East respiratory syndrome using an array of gold nanoparticle-modified carbon electrodes. Microchim Acta. https://doi.org/10.1007/s00604-019-3345-5

Lee E-Y, Ko HL, Jung S-Y, Nam J-H (2017) Development of a rapid diagnostic kit for antigen and antibody of Middle East respiratory syndrome coronavirus. J Immunol

Li D, Wang D, Dong J, Wang N, Huang H, Xu H, Xia C (2020a) False-negative results of real-time reverse-transcriptase polymerase chain reaction for severe acute respiratory syndrome coronavirus 2: role of deep-learning-based ct diagnosis and insights from two cases. Korean J Radiol. https://doi.org/10.3348/kjr.2020.0146

Li Z, Yi Y, Luo X, Xiong N, Liu Y, Li S, Sun R, Wang Y, Hu B, Chen W, Zhang Y, Wang J, Huang B, Lin Y, Yang J, Cai W, Wang X, Cheng J, Chen Z, Sun K, Pan W, Zhan Z, Chen L, Ye F (2020b) Development and clinical application of a rapid IgM-IgG combined antibody test for SARS-CoV-2 infection diagnosis. J Med Virol. https://doi.org/10.1002/jmv.25727

Lin YC, Dong SL, Yeh YH, Wu YS, Lan GY, Liu CM, Chu TC (2005) Emergency management and infection control in a radiology department during an outbreak of severe acute respiratory syndrome. Br J Radiol. https://doi.org/10.1259/bjr/17161223

Liu J, Chen X, Wang Q, Xiao M, Zhong D, Sun W, Zhang G, Zhang Z (2019) Ultrasensitive monolayer MoS 2 field-effect transistor based DNA sensors for screening of down syndrome. Nano Lett. https://doi.org/10.1021/acs.nanolett.8b03818

Liu C, Zhou Q, Li Y, Garner LV, Watkins SP, Carter LJ, Smoot J, Gregg AC, Daniels AD, Jervey S, Albaiu D (2020) Research and Development on therapeutic agents and vaccines for COVID-19 and related human coronavirus diseases. ACS Cent Sci. https://doi.org/10.1021/acscentsci.0c00272

Luo X, Davis JJ (2013) Electrical biosensors and the label free detection of protein disease biomarkers. Chem Soc Rev. https://doi.org/10.1039/c3cs60077g

Mahato K, Chandra P (2019) Paper-based miniaturized immunosensor for naked eye ALP detection based on digital image colorimetry integrated with smartphone. Biosens Bioelectron. https://doi.org/10.1016/j.bios.2018.12.006

Mahato K, Baranwal A, Srivastava A, Maurya PK, Chandra P (2018a) Smart materials for biosensing applications. In: Pawar PM, Ronge BP, Balasubramaniam R, Seshabhattar S (eds) Techno-societal 2016. Springer International Publishing, Cham, pp 421–431

Mahato K, Kumar S, Srivastava A, Maurya PK, Singh R, Chandra P (2018b) Electrochemical immunosensors: fundamentals and applications in clinical diagnostics. In: Handbook of immunoassay technologies. https://doi.org/10.1016/B978-0-12-811762-0.00014-1

Mahato K, Maurya PK, Chandra P (2018c) Fundamentals and commercial aspects of nanobiosensors in point-of-care clinical diagnostics. 3 Biotech. https://doi.org/10.1007/s13205-018-1148-8

Mahato K, Purohit B, Bhardwaj K, Jaiswal A, Chandra P (2019) Novel electrochemical biosensor for serotonin detection based on gold nanorattles decorated reduced graphene oxide in biological fluids and in vitro model. Biosens Bioelectron. https://doi.org/10.1016/j.bios.2019.111502

Mahato K, Purohit B, Kumar A, Chandra P (2020a) Paper based biosensors for clinical and biomedical applications: emerging engineering concepts and challenges. In: Comprehensive analytical chemistry. https://doi.org/10.1016/bs.coac.2020.02.001

Mahato K, Purohit B, Kumar A, Chandra P (2020b) Clinically comparable impedimetric immunosensor for serum alkaline phosphatase detection based on electrochemically engineered au-nano-Dendroids and graphene oxide nanocomposite. Biosens Bioelectron. https://doi.org/10.1016/j.bios.2019.111815

Matsuyama S, Nao N, Shirato K, Kawase M, Saito S, Takayama I, Nagata N, Sekizuka T, Katoh H, Kato F, Sakata M, Tahara M, Kutsuna S, Ohmagari N, Kuroda M, Suzuki T, Kageyama T, Takeda M (2020) Enhanced isolation of SARS-CoV-2 by TMPRSS2- expressing cells. Proc Natl Acad Sci U S A. https://doi.org/10.1073/pnas.2002589117

Mejía-Salazar JR, Oliveira ON (2018) Plasmonic biosensing. Chem Rev. https://doi.org/10.1021/acs.chemrev.8b00359

Memish ZA, Perlman S, Van Kerkhove MD, Zumla A (2020) Middle East respiratory syndrome. Lancet. https://doi.org/10.1016/S0140-6736(19)33221-0

Mohan R, Mach KE, Bercovici M, Pan Y, Dhulipala L, Wong PK, Liao JC (2011) Clinical validation of integrated nucleic acid and protein detection on an electrochemical biosensor array for urinary tract infection diagnosis. PLoS One. https://doi.org/10.1371/journal.pone.0026846

Nasir MU, Roberts J, Muller NL, Macri F, Mohammed MF, Akhlaghpoor S, Parker W, Eftekhari A, Rezaei S, Mayo J, Nicolaou S (2020) The role of emergency radiology in COVID-19: from preparedness to diagnosis. Can Assoc Radiol J. https://doi.org/10.1177/0846537120916419

Ozili PK, Arun T (2020) Spillover of COVID-19: impact on the global economy. SSRN Electron J. https://doi.org/10.2139/ssrn.3562570

Pan Y, Long L, Zhang D, Yan T, Cui S, Yang P, Wang Q, Ren S (2020) Potential false-negative nucleic acid testing results for severe acute respiratory syndrome coronavirus 2 from thermal inactivation of samples with low viral loads. Clin Chem. https://doi.org/10.1093/clinchem/hvaa091

Pejcic B, De Marco R, Parkinson G (2006) The role of biosensors in the detection of emerging infectious diseases. Analyst. https://doi.org/10.1039/b603402k

Polizzi KM (2019) Biosensors. In: Comprehensive biotechnology. https://doi.org/10.1016/B978-0-444-64046-8.00060-4

Prasad A, Mahato K, Maurya PK, Chandra P (2016) Biomaterials for biosensing applications. J Anal Bioanal Tech. https://doi.org/10.4172/2155-9872.1000e124

Purohit B, Kumar A, Mahato K, Chandra P (2019a) Novel sensing assembly comprising engineered gold dendrites and MWCNT-AuNPs Nanohybrid for acetaminophen detection in human urine. Electroanalysis. https://doi.org/10.1002/elan.201900551

Purohit B, Kumar A, Mahato K, Roy S, Chandra P (2019b) Cancer cytosensing approaches in miniaturized settings based on advanced nanomaterials and biosensors. In: Nanotechnology in modern animal biotechnology: concepts and applications. https://doi.org/10.1016/B978-0-12-818823-1.00009-0

Purohit B, Mahato K, Kumar A, Chandra P (2019c) Sputtering enhanced peroxidase like activity of a dendritic nanochip for amperometric determination of hydrogen peroxide in blood samples. Microchim Acta. https://doi.org/10.1007/s00604-019-3773-2

Purohit B, Kumar A, Mahato K, Chandra P (2020a) Smartphone-assisted personalized diagnostic devices and wearable sensors. Curr Opin Biomed Eng. https://doi.org/10.1016/j.cobme.2019.08.015

Purohit B, Kumar A, Mahato K, Chandra P (2020b) Electrodeposition of metallic nanostructures for biosensing applications in health care. J Sci Res. https://doi.org/10.37398/jsr.2020.640109

Qiu G, Gai Z, Tao Y, Schmitt J, Kullak-Ublick GA, Wang J (2020) Dual-functional Plasmonic Photothermal biosensors for highly accurate severe acute respiratory syndrome coronavirus 2 detection. ACS Nano. https://doi.org/10.1021/acsnano.0c02439

Rapp BE, Gruhl FJ, Länge K (2010) Biosensors with label-free detection designed for diagnostic applications. Anal Bioanal Chem. https://doi.org/10.1007/s00216-010-3906-2

Seo G, Lee G, Kim MJ, Baek S-H, Choi M, Ku KB, Lee C-S, Jun S, Park D, Kim HG, Kim S-J, Lee J-O, Kim BT, Park EC, Kim SI (2020) Rapid detection of COVID-19 causative virus (SARS-CoV-2) in human nasopharyngeal swab specimens using field-effect transistor-based biosensor. ACS Nano. https://doi.org/10.1021/acsnano.0c02823

Shen M, Zhou Y, Ye J, Abdullah AL-maskri AA, Kang Y, Zeng S, Cai S (2020) Recent advances and perspectives of nucleic acid detection for coronavirus. J Pharm Anal. https://doi.org/10.1016/j.jpha.2020.02.010

Sin ML, Mach KE, Wong PK, Liao JC (2014) Advances and challenges in biosensor-based diagnosis of infectious diseases. Expert Rev Mol Diagn. https://doi.org/10.1586/14737159.2014.888313

Vaddiraju S, Tomazos I, Burgess DJ, Jain FC, Papadimitrakopoulos F (2010) Emerging synergy between nanotechnology and implantable biosensors: a review. Biosens Bioelectron. https://doi.org/10.1016/j.bios.2009.12.001

Web Refereces 1. https://www.sinobiological.com/research/virus/2019-ncov-antigen

Wu HS, Chiu SC, Tseng TC, Lin SF, Lin JH, Hsu YF, Wang MC, Lin TL, Yang WZ, Ferng TL, Huang KH, Hsu LC, Lee LL, Yang JY, Chen HY, Su SP, Yang SY, Lin TH, Su IJ (2004) Serologic and molecular biologic methods for SARS-associated coronavirus infection. Taiwan Emerg Infect Dis. https://doi.org/10.3201/eid1002.030731

Wu D, Wu T, Liu Q, Yang Z (2020) The SARS-CoV-2 outbreak: what we know. Int J Infect Dis. https://doi.org/10.1016/j.ijid.2020.03.004

Xiao AT, Tong YX, Zhang S (2020) False-negative of RT-PCR and prolonged nucleic acid conversion in COVID-19: rather than recurrence. J Med Virol. https://doi.org/10.1002/jmv.25855

Yu J, Ding N, Chen H, Liu XJ, Pu ZH, Xu HJ, Lei Y, Zhang HW (2020) Loopholes in current infection control and prevention practices against COVID-19 in radiology department and improvement suggestions. Can Assoc Radiol J. https://doi.org/10.1177/0846537120916852

Zumla A, Hui DS, Perlman S (2015) Middle East respiratory syndrome. Lancet. https://doi.org/10.1016/S0140-6736(15)60454-8

Radiological Perspective of the Novel Coronavirus Disease 2019 (COVID-19)

Durgesh K. Dwivedi, Anit Parihar, Neera Kohli, Himanshu R. Dandu, and Shailendra K. Saxena ⓘ

Abstract

The novel Coronavirus Disease 2019 (COVID-19) pandemic has spread to more than 180 countries of the world. Chest imaging plays a critical role in screening and management of the disease. Chest X-ray is the most viable and economical radiological modality; however, it suffers from a lower sensitivity in the diagnosis of COVID-19. Therefore, CT is recommended for the screening of COVID-19. The predominant CT findings of COVID-19 infection are bilateral and peripheral ground-glass and consolidative pulmonary opacities. CT can be useful in assessing temporal changes in patients recovering from COVID-19. The knowledge about the disease is still evolving, and caution must be taken during the evaluation of chest CT of COVID-19. CT findings in children are also variable, but the most common findings are ground-glass opacities and consolidation. Radiology departments must implement strict infection control protocols. To minimize the COVID-19 spread, radiology departments should adopt team segregation strategies with minimum overlap of the personnel. This chapter discusses

Authors Durgesh K. Dwivedi, Shailendra K. Saxena have equally contributed to this chapter as first author.

D. K. Dwivedi (✉) · A. Parihar · N. Kohli
Department of Radiodiagnosis, King George's Medical University (KGMU), Lucknow, India
e-mail: durgeshdwivedi@kgmcindia.edu

H. R. Dandu
Department of Internal Medicine, King George's Medical University, Lucknow, India

S. K. Saxena
Centre for Advanced Research (CFAR), Faculty of Medicine, King George's Medical University (KGMU), Lucknow, India
e-mail: shailen@kgmcindia.edu

© The Editor(s) (if applicable) and The Author(s), under exclusive license to
Springer Nature Singapore Pte Ltd. 2020
P. Chandra, S. Roy (eds.), *Diagnostic Strategies for COVID-19 and other Coronaviruses*, Medical Virology: From Pathogenesis to Disease Control,
https://doi.org/10.1007/978-981-15-6006-4_3

the possible role of imaging methods and the recent advancement in key CT findings of COVID-19 infection, preparation of radiology departments, strategies to reduce the transmission, and personnel safety.

Keywords

COVID-19 · CT scan · Radiology · Chest · SARS-CoV-2 · Deep learning

1   Introduction

The novel Coronavirus Disease 2019 (COVID-19) outbreak, a highly infectious disease, is reported to have originated from Wuhan, Hubei province, China, in December, 2019 (Zhu et al. 2020b; Wu and McGoogan (2020)). There are several strains of coronaviruses; however, only two strains had been known to cause epidemics with high mortality rates – Middle East respiratory syndrome coronavirus (MERS-CoV) which originated in Saudi Arabia in 2012 and severe acute respiratory syndrome coronavirus (SARS-CoV) which originated in Guangdong province of China in 2002–2003 (Chan-Yeung and Xu 2003; Ramadan and Shaib 2019). The recent outbreak of COVID-19 has been caused by severe acute respiratory syndrome coronavirus2 (SARS-CoV-2) which is structurally related to the virus that causes severe acute respiratory syndrome (SARS). COVID-19 disease has spread globally, and its impact on public health system, on the financial status of vulnerable sections of population, and on enormous cost to the GDP of the nations is of great concern. The details of the current COVID-19 pandemic are still progressing. To date, no specific therapies of SARS-CoV-2 are available; however several groups are in the process of developing vaccines.

Patients with COVID-19 present with symptoms of lower respiratory tract infection, such as fever, dry cough, fatigue, myalgia, and dyspnea (Guan et al. 2020; Kanne et al. 2020). The real-time reverse transcription polymerase chain reaction (RT-PCR) of viral nucleic acid is regarded as the reference standard for the diagnosis of COVID-19. RT-PCR is highly specific, but with variable sensitivity of 60–97% (Kanne et al. 2020; Mossa-Basha et al. 2020). Given the highly contagious nature of COVID-19, rapid and reliable diagnosis of the disease is very important (Mahapatra and Chandra 2020). Chest imaging could play an important role in both the diagnosis and follow-up of COVID-19 (Guan et al. 2020). Further, chest CT could be advantageous where RT-PCR is false negative (Fang et al. 2020a, b; Xie et al. 2020). Imaging also helps in evaluating disease severity and progression of COVID-19.

2　　Strategy to Deal with COVID-19 Outbreak

Early data obtained from China, Italy, the UK, and the USA indicate that ethnicity might affect disease outcome, and among all nations, the USA has the largest number of COVID-19 cases (Khunti et al. 2020). The outcomes appear to be worse in patients having comorbidities (e.g., diabetes mellitus, hypertension, coronary artery disease) portending a higher morbidity and mortality among south Asians who have higher prevalence of these comorbidities (Tillin et al. 2013). As of April 30, 2020, a total of 33,049 cases have been registered by the Ministry of Health and Welfare in India, and the death rate is ~3.2% (https://www.mohfw.gov.in/index. php). Unlike the Germany and South Korea, where infected people were isolated based on excessive testing, India has adapted contact tracing and quarantine approach that could be relevant to other low-income countries facing kit shortages. Under a strict lockdown policy, the healthcare workers were instructed to trace and quarantine the infected people. Beside these, travel ban, high-level hand hygiene, social distancing, avoidance of social gathering, and the practice of yoga and meditation to reduce stress are being followed. In the medical centers, radiology departments have been instructed to train the entire teams after splitting them into smaller teams. All the nonessential procedures have been postponed. Mass screening of the patients is being performed at the hospital entrance, and patients are given a red or green tag in the holding area. Red tags are given to the COVID-19 suspected patients while green tag to the non-COVID patients. Dedicated mobile X-ray machines, CT, and ultrasound units have been provided to the COVID-19 wards.

3　　Imaging of COVID-19

Baseline chest X-ray is useful (Fig. 1) in those patients who have more severe forms of disease and require hospitalization where it is reported to have a sensitivity of 69% at the time of admission, compared to 91% for initial RT-PCR in COVID-19 patients (Wong et al. 2020). Chest CT is more sensitive than chest X-ray in COVID-19 patients. Although CT is highly sensitive in COVID-19, the appearance of COVID-19 pneumonia on the chest CT is thought to be nonspecific. In the early phase of the disease, CT typically shows bilateral multiple ground-glass opacity, with mostly subpleural distribution (Wu et al. 2020). A meta-analysis of computed tomography (CT) imaging features of COVID 19 reported that 74% of patients had bilateral lung involvement with 67% of patients having multilobar involvement on chest CT (Zhu et al. 2020a). About 8% of patients had no abnormality on CT. A study conducted on a total of 1099 COVI D-19 patients reported that on admission, 56% of patients had ground-glass opacities on chest CT (Guan et al. 2020). In severe cases, consolidation is reported and often superimposed on the ground-glass opacities. Pleural effusion, cavitation, CT halo sign, pneumothorax, and lymphadenopathy were rare in COVID-19 (Chen et al. 2020; Kong and Agarwal 2020; Salehi et al. 2020). Song et al. reported that isolated ground-glass opacity, ground glass opacity with reticular or interlobular septal thickening, and ground-glass opacity in combination with

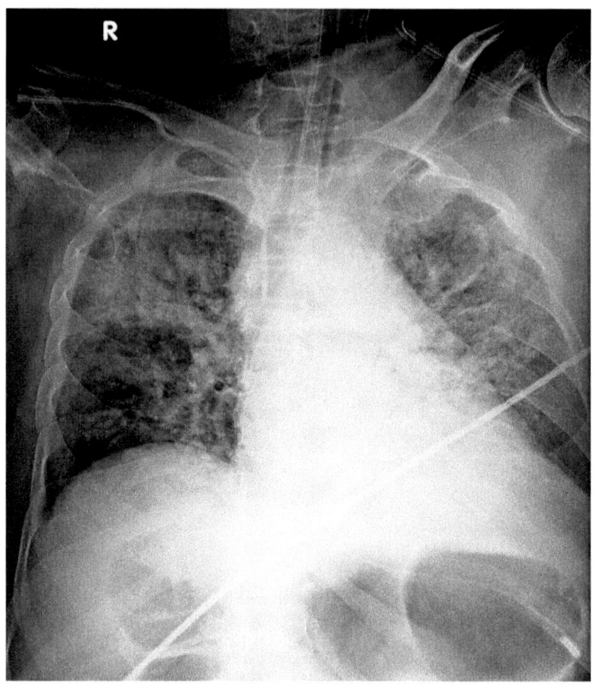

Fig. 1 A case of 70-year-old man with severe COVID-19 pneumonia. The patient had diabetes mellitus and hypertension. Portable chest X-ray shows bilateral ground-glass opacities involving all the lung zones

consolidative opacities were the most prominent pattern noted on chest CT (Song et al. 2020). They also observed a significant difference in pattern of chest CT based on the age group. The younger population (less than 50 years) had more ground-glass opacity, while the older age group (more than 50 years) had more areas of lung involvement and consolidation (Song et al. 2020).

CT findings in children were similar to those in adults. Bilateral lung involvement, ground-glass opacities, and consolidation were the common CT findings of COVID-19 in children (Xia et al. 2020). Further, lesions in children with COVID-19 were less diffuse than adults.

Although CT is a highly sensitive technique in COVID-19, the performance of radiologists in this disease may vary. A retrospective study evaluating the performance of radiologists in differentiating COVID-19 from viral pneumonia on chest CT showed that radiologists had moderate sensitivity but high specificity in distinguishing COVID-19 from viral pneumonia on chest CT (Bai et al. 2020). In another progress, the Radiological Society of North America (RSNA) has published an expert consensus statement on reporting chest CT findings related to COVID-19 (Simpson et al. 2020). This consensus statement will provide a consistent framework and confidence in reporting chest CT.

4   Radiological Differential Diagnosis of COVID-19 Pneumonia

Several studies have analyzed the utility of CT scan findings as surrogate marker for RT-PCR test for COVID-19. The current consensus is that CT findings cannot be used as replacement of RT-PCR for COVID-19 (Raptis et al. 2020). A number of infective and non-infective pathologies can simulate COVID-19 on CT scan. The notable differentials are non-COVID viral pneumonias and disorders like acute respiratory distress syndrome (ARDS) and organizing pneumonia which may occur during the course of any lung injury (Franquet 2011; Sheard et al. 2012; Baque-Juston et al. 2014).

4.1   Infective Causes

- SARS-Coronavirus (SARS-CoV) pneumonia – most commonly has unifocal opacity. Multifocal opacity when seen may have middle and lower lung predominance and may have unilateral or bilateral ground-glass opacity and crazy paving pattern (Paul et al. 2004; Franquet 2011).
- MERS-Corona virus pneumonia – closely resembles COVID-19 pneumonia radiographically with peripheral ground glass opacity and consolidation (Ajlan et al. 2014). Severe cases may show pneumothorax, pleural effusion, and progressive involvement of all the lung.
- Other seasonal viral pneumonia like a) influenza virus A (including avian flu (H5N1) and swine flu (H1N1) – they show more central involvement along the bronchovascular bundle with bronchial thickening (Koo et al. 2018). Influenza pneumonia may have nodules while avian flu may show pneumatoceles and pleural effusion.
- Parainfluenza 1–4-have multiple nodules along with ground-glass opacities and consolidations.
- Rhinovirus may closely resemble COVID-19 pneumonia as they show multiple patchy ground-glass opacities with interlobular thickening in both lungs.
- Adenovirus resembles bacterial pneumonia in that the bilateral multifocal ground-glass opacities and patchy consolidation show segmental or lobar distribution.
- Respiratory syncytial virus shows airway centric pattern, with tree-in-bud opacity and bronchial wall thickening
- Hanta virus may cause an outbreak of hantavirus pulmonary syndrome. It is characterized by more of central air space disease, prominence of interstitial edema, and early pleural effusion (Koo et al. 2018).
- Viral and nonviral pneumonias in predominantly immunocompromised patients – while the abovementioned viral pneumonias can occur in immunocompetent patients, few viruses like herpes, varicella, and CMV are more often encountered in immunocompromised patients (Nambu et al. 2014). These pneumonias often exhibit nodules. The immunocompromised status warrants keeping these in differential diagnosis.

- Bacterial pneumonia – clinical features and laboratory exams are often helpful in differentiating it from viral pneumonia (Koo et al. 2018). In consolidation, predominant pattern (alveolar/lobar pneumonia) is an air space consolidation in non-segmental distribution, e.g., *Streptococcus* and *Klebsiella* pneumonia. In peribronchial nodule-predominant pattern (bronchopneumonia), peribronchial nodules with or without consolidations is seen. Bronchial wall thickening, dilatation, mucus plugging, and centrilobular nodules are also seen. *Haemophilus influenzae*, mycoplasma, and *Chlamydophila* pneumonia are examples (Nambu et al. 2014).

4.2 Secondary to Lung Injury

ARDS and organizing pneumonia may occur secondary to lung damage by infection or due to non-infective causes like drugs, aspiration of gastric contents, vasculitis, etc. (Sheard et al. 2012; Baque-Juston et al. 2014). They may clinically resemble, confound, or complicate viral pneumonias. It is of great importance not only to differentiate between radiological appearances of COVID-19 pneumonia and ARDS and organizing pneumonia to increase accuracy of radiological diagnosis when RT-PCR is false negative or supply is constrained but also to recognize ARDS and complicating COVID-19 pneumonia, as this may entail change in therapy like prone position nursing, use and change of ventilator settings, and glucocorticoids.

1. ARDS – in the appropriate clinical context, ARDS is the presence of bilateral asymmetrical (pulmonary ARDS) or symmetrical (non-pulmonary ARDS), opacities with anteroposterior gradient. Normal or hyper-expanded lung anteriorly, ground-glass opacities are more posteriorly which condense into consolidations posteriorly. Bronchial dilatation is seen in ground glass opacities (Sheard et al. 2012). The similarity in clinical deterioration, time course, and radiological findings makes ARDS a close differential of COVID-19 pneumonia.
2. Organizing pneumonia – it occurs during repair process of lung injury and has a chronic course. It may superimpose on prior lung injury by viral pneumonia and other cause. It has many radiological patterns, but the classical form is multifocal often bilateral and symmetrical fluctuating subpleural and/or bronchovascular consolidations (Baque-Juston et al. 2014). These may have ground-glass opacities and traction bronchiectasis with air bronchogram sign. It may have lower lobe predominance. The migratory nature of lesions is considered to be reliable in differentiating organizing pneumonia from other mimics.

4.3 Non-infective Causes

1. Pulmonary edema – cardiomegaly, bilateral parahilar distribution, temporal changes, septal lines, and pleural effusion may help in differentiation (Sheard et al. 2012).

2. Pulmonary embolism and multiple pulmonary infarcts have been known to complicate COVID-19. CT angiography is required to confirm embolism.
3. Vasculitis – some vasculitis presents with opacities and ground-glass opacities (Polyangitis, Churg Strauss Syndrome), and some have diffuse alveolar hemorrhage (Goodpasture Syndrome, Henoch Schonlein Purpura, drug-induced vasculitis, SLE-vasculitis, IgA-nephropathy capilllaritis) (Chung et al. 2010). These radiological findings may mimic COVID-19 pneumonia, but the clinical presentation (multisystem at initial presentation) and serology tests like presence of certain autoantibodies like ANCA may help.
4. Drug-induced pneumonia – a number of drugs cause pneumonitis including immunotherapy agents. A case report of oseltamivir causing pneumonitis is also reported (Namba et al. 2006). In view of many drugs being tried in COVID-19 treatment, there is a need to be vigilant for drug-induced pneumonitis in such patients.
5. Acute eosinophilic pneumonia – ground-glass opacities and consolidation are seen. Most have pleural effusion and thickening of bronchovascular marking, and there may be history of new or binge smoking.

5   Imaging in Follow-Up of Patients with COVID-19 Infection

CT can be useful in assessing temporal changes in patients recovering from COVID-19. Pan et al. evaluated the changes in CT appearance in 21 patients of COVID-19 from initial diagnosis until patient recovery (Pan et al. 2020). Repeat chest CT was performed, and a total of 82 pulmonary CT scans were acquired with a mean interval of 4 ± 1 days (range: 1–8 days). Four stages of CT were used to assess the severity; in the early stage, till day 4, 17% of patients showed no lung abnormalities, 75% of patients showed ground glass opacities, and 42% of patients showed consolidation. CT scan at 5–8 days revealed 82% of patients showed ground-glass opacities that extended to more pulmonary lobes with more of crazy-paving pattern in 53% of patients. Consolidation was the most common features in 91% of patients between 9 and 13 days, and this period was referred as the peak stage (Pan et al. 2020). Also, CT findings were more pronounced at 10–12 days. Absorption stage started at 14 days after the onset of initial symptoms. After 14 days, 75% of patients showed improvement without any crazy paving pattern and consolidation. Others have shown that follow-up chest CT showed increased consolidation with loss of the areas of crazy paving. Further progressive pulmonary opacities, progression to a mixed pattern of ground-glass opacities, and development of pleural effusions have been reported on follow-up CT (Fang et al. 2020a, b; Lei et al. 2020).

6 Deep Learning Approaches in COVID-19 Using Chest CT

Recently, deep learning with convolutional neural networks (CNNs) has revolutionized the automated analysis of medical images (Fig. 2) (Shen et al. 2017). A few studies have shown the potential of deep learning for the accurate diagnosis of COVID-19 that saves critical time for disease control (Huang et al. 2020; Li et al. 2020).

Li et al. developed a three-dimensional deep learning model, COVID-19 detection neural network (COVNet) for the detection of COVID-19 on chest CT. The model was used to distinguish COVID-19 from community-acquired pneumonia. The model achieved a very high sensitivity of 90% and specificity of 96% in the detection of COVID-19. The area under the receiver operating curve (AUC) for COVID-19 was 0.96, while the AUC for community-acquired pneumonia was 0.95. The study showed that the deep learning model can differentiate COVID-19 from community-acquired pneumonia and other lung diseases (Li et al. 2020). Huang et al. evaluated the longitudinal changes in COVID-19 at baseline and follow-up imaging (Huang et al. 2020). They observed that the deep learning-based quantitative CT parameters (opacification percentage) can provide an insight into the severity of the pulmonary manifestations of COVID-19 and monitoring disease progression (Huang et al. 2020). Further larger studies are needed to determine the performance of deep learning approaches in COVID-19 on chest CT.

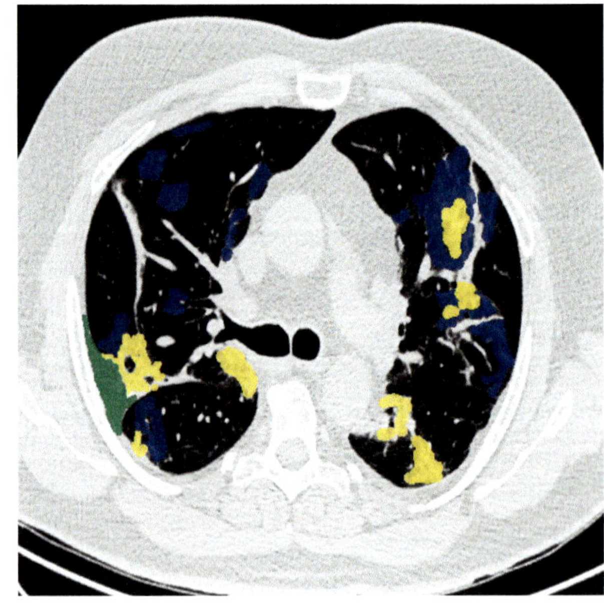

Fig. 2 Representative example of AI-based segmentation of COVID-19 CT axial slice. The authors took the case example from, an open-source dataset, the Italian Society of Medical and Interventional Radiology (SIRM, https://www.sirm.org/category/senza-categoria/covid-19/) for AI-based segmentation. Ground-glass opacities are shown in blue, consolidation in yellow, and pleural effusion in green. (Source: adapted from http://medicalsegmentation.com/covid19/; reprinted with permission from SIRM, secretariat)

7   Protection of Radiology Personnel

Evidence suggests that the 3.8% of healthcare worker were infected with SARS-CoV-2 in China (Wu et al. 2020). Therefore, the safety of healthcare workers including radiology personnel is paramount. All the staffs must be given a proper safety training to get a detailed knowledge of COVID-19 through e-learning platforms in small groups. It is better to provide surgical masks to all patients and staff at the hospital entrance. Majority of centers have started screening patients at the main entrance with an additional level of screening at the radiology front. Mossa-Basha et al. detailed the guidelines adapted at the University of Washington for COVID-19 (Mossa-Basha et al. 2020). Most of the elective diagnostic imaging and other nonessential procedures have been rescheduled to protect the patients and employees from COVID-19 exposure. At this center, imaging is not routinely performed for COVID-19 screening but to rule out other diagnoses that can be treated, including pulmonary embolism. Portable radiography and/or a dedicated CT scanner for COVID-19 patients can be helpful (Fig. 3). Donning and doffing the personal protective equipment (PPE) should be done in the designated area in a proper sequence (John et al. 2020). Fit-tested N95 filtering facepiece respirators can be used for droplet protection. A proper room decontamination protocol must be followed scanning COVID-19 patients. X-ray exposure through a glass window has been more effective in which nurse holds a double-bagged X-ray cassette behind the patient. The nurse disinfects the outer bag, while technologist disinfects the inner bag for further processing of the film (Mossa-Basha et al. 2020).

Quaternary ammonium/alcohol impregnated wipes or Environmental Protection Agency (EPA)-approved disinfectants can be used for the equipment sanitization (Mossa-Basha et al. 2020). Diluted bleach solution can be used to clean the exterior of the portable X-ray. Seventy percent isopropyl alcohol should be used to clean the delicate parts such as collimators and exposure buttons of the radiology equipment. Other indoor areas should be mopped with disinfectant with 1% sodium hypochlorite or phenolic disinfectants. It is recommended to shut down the imaging suites for 1 h for airborne precautions after scanning a COVID-19 case. A negative pressure environment should be setup to perform interventional radiology procedures. To minimize the COVID-19 exposure, radiology department should adopt team segregation strategies with minimum overlap of the personnel. Further radiology personnel must be on rotation with 1 week on and the following week off, and the remote support should be available by the off team.

8   Summary

- Chest CT can show no abnormality in a few subsets of patients in the very early onset of COVID-19.
- COVID-19 RT-PCR is highly specific but with low sensitivity. CT could be useful in patients with negative RT-PCR.

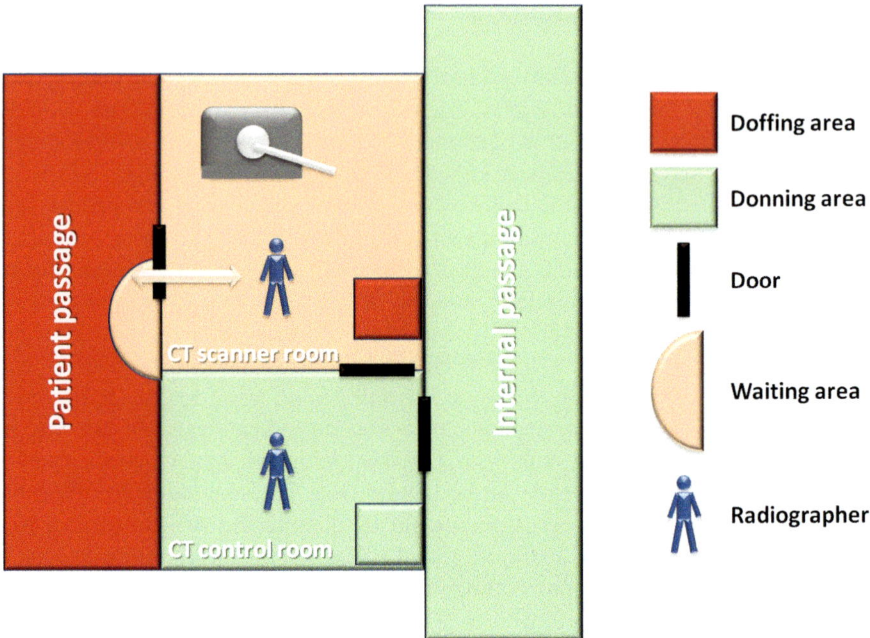

Fig. 3 Schematic of the dedicated CT scan setup for COVID-19. It is recommended that two radiographers perform the duty. One remains in the CT control room to acquire CT scan of COVID-19 infection. The other radiographer dons full personal protective equipment (PPE) and prepares and positions the patient and sanitizes the equipment in the CT scanner room. After cleaning the room, radiographer doffs the PPE at the designated area and performs hand hygiene before leaving the CT room. The standard acquisition parameters for a 64-slice non-contrast CT scan: patient in the supine position, end inspiratory acquisition; 120 kVp with automatic tube current modulation or 100–200 mAs, and section thickness after reconstruction 1.5 or 1 mm. Here, red represents contaminated area, orange represents a mixed zone, and green represents a clean area

- Majority of patients had bilateral, multifocal lower lung involvement with subpleural/peripheral.
- Ground-glass opacity, mixed ground-glass opacity, and ground-glass opacity in combination with consolidative opacities were the most prominent pattern observed on chest CT.
- CT plays a very important role in assessing the progression of COVID-19. As the disease progresses, consolidation and crazy-paving pattern were the most common features which peaked between 9 and 13 days.
- Portable imaging equipment should be preferred, and strict infection control must be on place.
- It is recommended that radiology departments must prepare to deal with patient influx by limiting the elective imaging and procedures and standardized approaches, which would reduce exposures to radiology personnel. Further, radiology department must adopt team segregation strategies to minimize the COVID-19 exposure.

9   Conclusions

In conclusion, chest imaging plays an important role in the management of COVID-19 patients. The most common CT imaging features in COVID-19 are bilateral and peripheral ground-glass and consolidative pulmonary opacities. CT is recommended for follow-up that will help in assessing the recovery of patient from COVID-19. Serial CT imaging shows progression of COVID-19 with consolidation and crazy-paving pattern as the prominent features.

10   Future Perspective

CT is recommended for follow-up of patients with COVID-19 infection. New imaging patterns may emerge as larger studies are conducted. Though portable CT is expensive, it can provide a potential solution in the screening of COVID-19. Future studies can help decide how patients with parenchymal lung disease evolve after treatment. Further to automate the diagnosis of COVID-19, larger studies are required to determine the performance of deep learning algorithms in COVID-19 on chest X-ray or CT.

Acknowledgments The authors are grateful to the Vice Chancellor, King George's Medical University (KGMU), Lucknow, India, for the encouragement of this work. The authors have no other relevant affiliations or financial involvement with any organization or entity with a financial interest in or financial conflict with the subject matter or materials discussed in the manuscript apart from those disclosed.

References

Ajlan AM, Ahyad RA, Jamjoom LG, Alharthy A, Madani TA (2014) Middle East respiratory syndrome coronavirus (MERS-CoV) infection: chest CT findings. AJR Am J Roentgenol 203 (4):782–787

Bai HX, Hsieh B, Xiong Z, Halsey K, Choi JW, Tran TML, Pan I, Shi LB, Wang DC, Mei J, Jiang XL, Zeng QH, Egglin TK, Hu PF, Agarwal S, Xie FF, Li S, Healey T, Atalay MK, Liao WH (2020) Performance of radiologists in differentiating COVID-19 from non-COVID-19 viral pneumonia at chest CT. Radiology 296(2):E46–E54

Baque-Juston M, Pellegrin A, Leroy S, Marquette CH, Padovani B (2014) Organizing pneumonia: what is it? A conceptual approach and pictorial review. Diagn Interv Imag 95(9):771–777

Chan-Yeung M, Xu RH (2003) SARS: epidemiology. Respirology 8(Suppl):S9–S14

Chen N, Zhou M, Dong X, Qu J, Gong F, Han Y, Qiu Y, Wang J, Liu Y, Wei Y, Xia J, Yu T, Zhang X, Zhang L (2020) Epidemiological and clinical characteristics of 99 cases of 2019 novel coronavirus pneumonia in Wuhan, China: a descriptive study. Lancet 395(10223):507–513

Chung MP, Yi CA, Lee HY, Han J, Lee KS (2010) Imaging of pulmonary vasculitis. Radiology 255 (2):322–341

Fang Y, Zhang H, Xie J, Lin M, Ying L, Pang P, Ji W (2020a) Sensitivity of chest CT for COVID-19. comparison to RT-PCR. Radiology 296(2).E115–E117

Fang Y, Zhang H, Xu Y, Xie J, Pang P, Ji W (2020b) CT manifestations of two cases of 2019 novel coronavirus (2019-nCoV) pneumonia. Radiology 295(1):208–209

Franquet T (2011) Imaging of pulmonary viral pneumonia. Radiology 260(1):18–39

Guan WJ, Ni ZY, Hu Y, Liang WH, Ou CQ, He JX, Liu L, Shan H, Lei CL, DSC H, Du B, Li LJ, Zeng G, Yuen KY, Chen RC, Tang CL, Wang T, Chen PY, Xiang J, Li SY, Wang JL, Liang ZJ, Peng YX, Wei L, Liu Y, Hu YH, Peng P, Wang JM, Liu JY, Chen Z, Li G, Zheng ZJ, Qiu SQ, Luo J, Ye CJ, Zhu SY, Zhong NS, China Medical Treatment Expert Group for C (2020) Clinical characteristics of Coronavirus disease 2019 in China. N Engl J Med 382(18):1708–1720

Huang L, Han R, Ai T, Yu P, Kang H, Tao Q, Xia L (2020) Serial quantitative chest CT assessment of COVID-19: deep-learning approach. Radiol Cardiothoracic Imag 2(2):e200075

John TJ, Hassan K, Weich H (2020) Donning and doffing of personal protective equipment (PPE) for angiography during the COVID-19 crisis. Eur Heart J 41:1786–1787

Kanne JP, Little BP, Chung JH, Elicker BM, Ketai LH (2020) Essentials for radiologists on COVID-19: an update-radiology scientific expert panel. Radiology 296(2):E113–E114

Khunti K, Singh AK, Pareek M, Hanif W (2020) Is ethnicity linked to incidence or outcomes of covid-19? BMJ 369:m1548

Kong W, Agarwal PP (2020) Chest imaging appearance of COVID-19 infection. Radiol Cardiothoracic Imag 2(1):e200028

Koo HJ, Lim S, Choe J, Choi SH, Sung H, Do KH (2018) Radiographic and CT features of viral pneumonia. Radiographics 38(3):719–739

Lei J, Li J, Li X, Qi X (2020) CT imaging of the 2019 Novel Coronavirus (2019-nCoV) pneumonia. Radiology 295(1):18–18

Li L, Qin L, Xu Z, Yin Y, Wang X, Kong B, Bai J, Lu Y, Fang Z, Song Q, Cao K, Liu D, Wang G, Xu Q, Fang X, Zhang S, Xia J, Xia J (2020) Using artificial intelligence to detect COVID-19 and community-acquired pneumonia based on pulmonary CT: evaluation of the diagnostic accuracy. Radiology 296(2):E65–E71

Mahapatra S, Chandra P (2020) Clinically practiced and commercially viable nanobio engineered analytical methods for COVID-19 diagnosis: Biosens. Bioelectron 165:112361

Mossa-Basha M, Medverd J, Linnau K, Lynch JB, Wener MH, Kicska G, Staiger T, Sahani D (2020) Policies and guidelines for COVID-19 preparedness: experiences from the university of Washington. Radiology 8:201326

Namba Y, Kitada S, Mori M, Iwasaki T, Motone M, Niinaka M, Yoshimura K, Miki M, Miki K, Naka N, Hiraga T, Maekura R, Ito M (2006) A case of pneumonitis suspected to be induced by Oseltamivir. Nihon Kokyuki Gakkai Zasshi 44(5):410–414

Nambu A, Ozawa K, Kobayashi N, Tago M (2014) Imaging of community-acquired pneumonia: roles of imaging examinations, imaging diagnosis of specific pathogens and discrimination from noninfectious diseases. World J Radiol 6(10):779–793

Pan F, Ye T, Sun P, Gui S, Liang B, Li L, Zheng D, Wang J, Hesketh RL, Yang L, Zheng C (2020) Time course of lung changes at chest CT during recovery from coronavirus disease 2019 (COVID-19). Radiology 295(3):715–721

Paul NS, Roberts H, Butany J, Chung T, Gold W, Mehta S, Konen E, Rao A, Provost Y, Hong HH, Zelovitsky L, Weisbrod GL (2004) Radiologic pattern of disease in patients with severe acute respiratory syndrome: the Toronto experience. Radiographics 24(2):553–563

Ramadan N, Shaib H (2019) Middle East respiratory syndrome coronavirus (MERS-CoV): a review. Germs 9(1):35–42

Raptis CA, Hammer MM, Short RG, Shah A, Bhalla S, Bierhals AJ, Filev PD, Hope MD, Jeudy J, Kligerman SJ, Henry TS (2020) Chest CT and coronavirus disease (COVID-19): a critical review of the literature to date. AJR Am J Roentgenol 215(4):839–842

Salehi S, Abedi A, Balakrishnan S, Gholamrezanezhad A (2020) Coronavirus disease 2019 (COVID-19): a systematic review of imaging findings in 919 patients. AJR Am J Roentgenol 215(1):87–93

Sheard S, Rao P, Devaraj A (2012) Imaging of acute respiratory distress syndrome. Respir Care 57 (4):607–612

Shen D, Wu G, Suk HI (2017) Deep learning in medical image analysis. Annu Rev Biomed Eng 19:221–248

Simpson S, Kay FU, Abbara S, Bhalla S, Chung JH, Chung M, Henry TS, Kanne JP, Kligerman S, Ko JP, Litt H (2020) Radiological Society of North America expert consensus statement on reporting chest CT findings related to COVID-19. Endorsed by the Society of Thoracic Radiology, the American College of Radiology, and RSNA. Radiol Cardiothoracic Imag 2(2): e200152

Song F, Shi N, Shan F, Zhang Z, Shen J, Lu H, Ling Y, Jiang Y, Shi Y (2020) Emerging 2019 Novel Coronavirus (2019-nCoV) pneumonia. Radiology 295(1):210–217

Tillin T, Hughes AD, Mayet J, Whincup P, Sattar N, Forouhi NG, McKeigue PM, Chaturvedi N (2013) The relationship between metabolic risk factors and incident cardiovascular disease in Europeans, South Asians, and African Caribbeans: SABRE (Southall and Brent revisited) – a prospective population-based study. J Am Coll Cardiol 61(17):1777–1786

Wong HYF, Lam HYS, Fong AH, Leung ST, Chin TW, Lo CSY, Lui MM, Lee JCY, Chiu KW, Chung TW, Lee EYP, Wan EYF, Hung IFN, Lam TPW, Kuo MD, Ng MY (2020) Frequency and distribution of chest radiographic findings in patients positive for COVID-19. Radiology 296(2):E72–E78

Wu Z, McGoogan JM (2020) Characteristics of and important lessons from the Coronavirus disease 2019 (COVID-19) outbreak in China: summary of a report of 72314 cases from the Chinese Center for Disease Control and Prevention. JAMA 323:1239–1242

Wu J, Wu X, Zeng W, Guo D, Fang Z, Chen L, Huang H, Li C (2020) Chest CT findings in patients with Coronavirus disease 2019 and its relationship with clinical features. Investig Radiol 55 (5):257–261

Xia W, Shao J, Guo Y, Peng X, Li Z, Hu D (2020) Clinical and CT features in pediatric patients with COVID-19 infection: different points from adults. Pediatr Pulmonol 55(5):1169–1174

Xie X, Zhong Z, Zhao W, Zheng C, Wang F, Liu J (2020) Chest CT for typical coronavirus disease 2019 (COVID 19) pneumonia: relationship to negative RT PCR testing. Radiology 296(2): E41–E45

Zhu J, Zhong Z, Li H, Ji P, Pang J, Li B, Zhang J (2020a) CT imaging features of 4121 patients with COVID-19: a meta-analysis. J Med Virol 92(7):891–902

Zhu N, Zhang D, Wang W, Li X, Yang B, Song J, Zhao X, Huang B, Shi W, Lu R, Niu P, Zhan F, Ma X, Wang D, Xu W, Wu G, Gao GF, Tan W, China Novel Coronavirus I, Research T (2020b) A Novel Coronavirus from patients with pneumonia in China, 2019. N Engl J Med 382 (8):727–733

Differential Diagnosis and Possible Therapeutics for Coronavirus Disease 2019

Anjani Devi Chintagunta, Mahesh Kumar, N. S. Sampath Kumar, and S. P. Jeevan Kumar

Abstract

Coronavirus disease 2019 (COVID-19) has now become a pandemic and major risk to world's health and economy. COVID-19 is caused by a viral pathogen named as severe acute respiratory syndrome coronavirus-2 (SARS-CoV-2). Scientists are working on various possible therapeutic agents to combat COVID-19. Based on the mechanism of SARS-CoV-2 pathogenesis, numbers of agents have been proposed for the treatment, but unfortunately, none of them have been approved for treating COVID-19, till date. To combat this pandemic, identification of SARS-CoV-2 patients is paramount, and because its symptoms are similar to the patients of other respiratory infections, differential diagnosis becomes very important. Differential diagnosis distinguishes a particular disease or condition from others that present similar clinical features. This book chapter deals with differential diagnosis in COVID-19 and current knowledge of the possible therapeutics including herbal- or plant-based agents and plasma therapy for the treatment of COVID-19.

Authors Anjani Devi Chintagunta, Mahesh Kumar have equally contributed to this chapter.

A. D. Chintagunta · N. S. Sampath Kumar
Department of Biotechnology, Vignan's Foundation for Science, Technology and Research, Vadlamudi, India

M. Kumar (✉)
Department of Biochemistry, College of Agriculture, Central Agricultural University, Pasighat, India

S. P. Jeevan Kumar
Department of Seed Biotechnology, ICAR-Indian Institute of Seed Science, Mau, India

P. Chandra, S. Roy (eds.), *Diagnostic Strategies for COVID-19 and other Coronaviruses*, Medical Virology: From Pathogenesis to Disease Control, https://doi.org/10.1007/978-981-15-6006-4_4

Keywords

Differential diagnosis · COVID-19 · Herbal therapeutics · Plasma therapy · SARS-CoV-2

1 Introduction

Human coronaviruses have not given much attention and are considered insignificant pathogens, which causes the "common cold" in otherwise healthy person (Paules et al. 2020). However, attention has been turned toward coronaviruses owing to the outbreak in the twenty-first century by two highly pathogenic human coronaviruses such as Middle East respiratory syndrome coronavirus (MERS-CoV) and severe acute respiratory syndrome coronavirus (SARS-CoV), respectively.

Coronaviruses were appeared from animal reservoirs and caused global epidemics with frightening illness and mortality (Paules et al. 2020; Guarner 2020; Peeri et al. 2020). Recently, in December 2019, a novel contagious human coronavirus outbreak happened in Wuhan city of Hubei Province in China; 27 patients were identified to suffer from pneumonia of unknown reason (Paules et al. 2020; Guarner 2020; Peeri et al. 2020; Zhu et al. 2020; Lu et al. 2020b). These all cases were connected to Huanan Seafood Market of Wuhan known for selling of different live animal species including bats (Lu et al. 2020b; Zhu et al. 2020). The causative pathogen of this disease was confirmed on 7th January 2020 and is referred as novel coronavirus 2019 (2019-nCoV) (Guarner 2020; Sohrabi et al. 2020), which was officially termed as severe acute respiratory syndrome coronavirus-2 (SARS-CoV-2) by the World Health Organization (WHO) on 11th February 2020 (World Health Organisation 2020a). The disease was named as coronavirus disease 2019 (COVID-19) (World Health Organisation 2020a).

SARS-CoV-2 belongs to subfamily *Coronavirinae* and family *Coronaviridae* (Zhu et al. 2020; Chan et al. 2020a). It is a *Betacoronavirus* comprising a single-stranded ribonucleic acid (RNA) genome (Zhu et al. 2020; Chan et al. 2020a). Phylogenetic study showed close relation (88–89% similarity) of SARS-CoV-2 with bat-SL-CoVZC45 and bat- SL-CoVZXC21SARS-like coronaviruses of bat-origin, but showed around 50% and 79% similarity with MERS-CoV and SARS-CoV coronaviruses, respectively (Zhu et al. 2020; Ren et al. 2020; Jiang et al. 2020; Lai et al. 2020). The virus SARS-COV-2 is highly infectious and transmissible much more than the other coronaviruses SARS-CoV and MERS-CoV. It quickly spreads in China and then in other parts of the world and is soon declared as a pandemic on 11th March 2020 (World Health Organisation 2020b). As of 2 May 2020, COVID-19 has been spread in 215 countries in the world with 32,78,272 confirmed cases and 2,36,457 deaths (WHO Coronavirus Disease (COVID-19) Dashboard 2020).

Although initial report suggested SARS-CoV-2 transmission from animal to human, soon it was evident that the virus can be transmitted from one human to another (Li et al. 2020b; Chang et al. 2020a; Wang et al. 2020a). Person-to-person transmission occurs due to close contact with the virus-infected person upon direct exposure to respiratory droplets and indirect exposure with surfaces in immediate contacts with the infected person (Liu et al. 2020; Chan et al. 2020b; World Health Organisation 2020c; Centers for Disease Control and Prevention 2020a). The incubation period for the disease COVID-19 is on average 5–6 days and however can be up to 14 days (World Health Organisation 2020c), while Li et al. found that exhibition of symptoms by the patients occurs on about 12.5 days from the time of infection (Li et al. 2020b).

Usually, respiratory viruses are most contagious during symptomatic period. However, it is evident from studies that COVID-19 spreading from one human to another may also be taking place during asymptomatic period (Rothe et al. 2020; Ye et al. 2020; Li et al. 2020b). Reports so far indicated that while patients of all ages are susceptible to the disease, older person and person with chronic or preceding medical condition/s are more vulnerable (Jordan et al. 2020; Centers for Disease Control and Prevention 2020b). COVID-19 patients with comorbid conditions showed higher mortality rates and poor recoveries (Jordan et al. 2020; Zhou et al. 2020a; Yang et al. 2020a). Health agencies have advocated that the COVID-19 spreading chain may be broken by timely detection, isolation, appropriate treatment, and contact tracing (World Health Organisation 2020d; European Centre for Disease Prevention and Control 2020; Sohrabi et al. 2020). Various health agencies have issued guidelines such as social distancing, avoid traveling to high risk area, regular washing of hands, and usage of PPE face masks to prevent further spread of COVID-19 (Centers for Disease Control and Prevention 2020a; Rothe et al. 2020; World Health Organisation 2020e).

Studies indicated that most of the COVID-19 patients showed mild initial symptoms such as fever, dry cough, sore throat, and shortness of breath (Guarner 2020; Chen et al. 2020a, b; Sohrabi et al. 2020), while some have developed serious complications like pulmonary edema, acute respiratory distress syndrome (ARDS), severe pneumonia, septic shock, organ failure, etc. (Chen et al. 2020b). The initial symptoms exhibited by COVID-19 patients are similar to the symptoms of patients with other respiratory infections (Singhal 2020; Chhikara et al. 2020), and hence differential diagnosis is very important in order to identify COVID-19 patients.

The distinction of COVID-19 from other respiratory infections is crucial for proper treatment. Currently, there is no approved therapeutics for COVID-19 (Chhikara et al. 2020; Zhang and Liu 2020; Ziaie et al. 2020). However, a number of therapeutics are in use or can possibly be utilized for the treatment of the disease based on the knowledge of similar kind of infections, mechanism of the SARS-COV-2 infection, etc. Recently, on 28th March 2020, emergency use of chloroquine phosphate and hydroxychloroquine sulfate oral formulations was approved by Food and Drug administration (FDA) for treating COVID-19 patients (U.S Food & Drug Administration 2020). Various countries are recommended to use a number of therapeutics for treating patients with this pandemic disease. In this book chapter, we are discussing the differential diagnosis and current knowledge on possible therapeutics for COVID-19.

2 Differential Diagnosis for COVID-19

Differential diagnosis distinguishes a particular disease or condition from others that present similar clinical features. The symptoms exhibited by patients with SARS-CoV infection are not specific (Guarner 2020; Sohrabi et al. 2020; Chen et al. 2020b) but similar to other respiratory infections like influenza, adenovirus, coronavirus other than COVID-19, parainfluenza, mycoplasma infection, bacterial infections, etc. and hence cannot be used for an accurate diagnosis (Singhal 2020). This made the differential diagnosis significant for COVID-19 treatment. The differentiation of COVID-19 from other respiratory infections is not possible through routine lab tests (Singhal 2020). Travel and exposure history of the suspected COVID-19 person is suggested to be crucial while differentiating from other respiratory infections (Singhal 2020; Wong et al. 2020).

Molecular techniques are considered to be more suitable diagnostic to accurately diagnose COVID-19 and differentiate it with other respiratory infections (Udugama et al. 2020; Mahapatra and Chandra 2020) because they target specific and precisely detect the particular pathogen (Udugama et al. 2020). Many kits pertinent to reverse transcription polymerase chain reaction (RT-PCR) have been developed and in use for detecting SARS-CoV-2 (Udugama et al. 2020). RT-PCR of COVID-19 using respiratory samples is the most predominant method of diagnosis (World Health Organisation 2020f;g). The samples for the test were collected either from upper or lower respiratory tract (suggested for patients showing wet cough) (Udugama et al. 2020). Nasopharyngeal swabs, nasopharyngeal washes, oropharyngeal swabs, and nasal aspirates are the upper respiratory samples that are mostly recommended (Udugama et al. 2020). The lower respiratory tract samples include sputum, bronchoalveolar lavage (BAL) fluid, and tracheal aspirates (Udugama et al. 2020). The detectable viral load is influenced by number of days after the onset of illness.

The most consistent detection of SARS-COV-2 could be observed in sputum and then by nasal swab in the first 14 days after symptom onset, while 8 days onwards after symptom onset, throat swab was found to be unreliable (Yang et al. 2020c; Pan et al. 2020). RT-PCR test is not 100% accurate and can give false-negative and false-positive results (Tahamtan and Ardebili 2020). False-positive results are generally due to sample contamination, while low viral load, improper sample collection, and mutation in the viral genome lead into false-negative results (Tahamtan and Ardebili 2020; Ai et al. 2020; Winichakoon et al. 2020). Winichakoon et al. presented an evidence for positive COVID-19 patients although the results from nasopharyngeal and/or oropharyngeal swab are negative (Winichakoon et al. 2020). Li and Xia found that chest computed tomography (CT) may be a worthwhile method for rapid detection of SARS-COV-2 infections; however, it is still limited to identify specific virus and differentiate between viruses (Li and Xia 2020).

Lei et al. (2020) suggested five relevant factors for differential diagnosis of COVID-19, which includes RT-PCR test results, CT findings, exposure history, laboratory test results, and clinical manifestations of SARS-CoV-2 infection (Lei et al. 2020). They further opined that the suspected cases can be grouped into five categories based on the information about all the mentioned factors: first is "definitely COVID-19," which will include patients with positive result of RT-PCR for SARS-COV-2 irrespective of other results; second is "very probably COVID-19," this will include patients with negative RT-PCR results but positive results for other relevant factors; third is "probably COVID-19," patients with a negative initial RT-PCR test result but having exposure history to an infected person and positive chest CT findings will be considered in this group; fourth is "probably not COVID-19," in this group patients have exposure history to an infected person but negative RT-PCR and chest CT findings will be included; and fifth is "very probably not COVID-19," patients with exposure history to an infected person but having negative results for all other factors will be considered in this group (Lei et al. 2020).

3　Possible Therapeutics for COVID-19 Treatment

The outbreak of SARS-CoV-2 resulted in coronavirus disease (COVID-19), which causes fatal respiratory disease due to aggressive inflammation of lung AT2 alveolar epithelial cells (Guan et al. 2020). Viral replications inside the host body exhibit immune response by increasing the levels of IL-1b, IFN-c, IP-10, MCP-1, IL-4, and IL-10 suggesting the severity of the disease (Huang et al. 2020). The existing drugs were partially successful in managing the symptoms, but no defined line of treatment or therapeutics is available till date because the pathophysiology of SARS-CoV-2 is not known fully (Chen et al. 2020b). So most of the drugs used in the line of action by doctors in controlling the COVID-19 are based on the symptoms and its similarities with SARS-CoV (Fu et al. 2020). Considering the mechanism of viral infection, starting with invasion of virus in the cells and its multiplication using host cellular machinery along with damage of the host cells, is the key part to look forward for the development of therapeutics. In view of this, the disease can be controlled by (i) blocking at its entry point (entry inhibitor), (ii) inhibiting its multiplication (viral protease inhibitor/viral replication inhibitor/ heterocyclic antiviral drugs), and (iii) improving immunity of host to reduce/replace the damaged cells (biological therapeutics/herbal therapeutics).

3.1　Entry Inhibitors

The SARS-coronavirus receptor utilizes angiotensin-converting enzyme 2 (ACE2) and the cellular transmembrane protease serine 2 (TMPRSS2) to enter target cells. External body organs (facial and oral) with a high density of these proteins are particularly vulnerable to the novel coronavirus and thus can become entry gates for COVID-19. High affinity of SARS-CoV-2 for ACE2 is prime reason behind its rapid transmission among human beings and according to the preclinical study; degree of infectivity of coronavirus depends upon the degree of ACE2 expression (Li et al. 2003). Thus, the impact of the viral infection can be reduced by implementing the strategies that decrease the expression of ACE2. It is hypothesized that endogenous mammalian peptide angiotensin II (AngII) prevents the viral infection in several ways.

I. AngII normally binds to ACE2 and hydrolysis into angiotensin-(1–7) and competes with SARS-CoV-2 for the receptor (Chawla et al. 2018).
II. Binding of AngII to the angiotensin II receptor type 1(AT1) causes internalization and downregulation of ACE2 through an ERK1/2 (extracellular signal-regulated protein kinase) and p38 MAP (mitogen-activated protein) kinase pathway (Fernandes et al. 2011).
III. AngII causes AT1 receptor-dependent destruction of ACE2 through ubiquitination and transport into lysosomes.

These actions may potentially prevent SARS-CoV-2 from entering the cell. Some investigators hypothesized that the expression of ACE2 increases in the presence of ACE inhibitors and angiotensin receptor blockers (ARBs) (Ferrario et al. 2005; Ishiyama et al. 2004). COVID-19 disease was found to be severe in the patients with renin-angiotensin-aldosterone system (RAAS) blockade therapy, hypertension, and diabetes mellitus (Guan et al. 2020). Drugs should be developed to block virus to integrate with ACE2, to activate the ACE2-Ang-(1–7)-Mas receptor pathway, or to inhibit ACE-AngII-AT1R receptor pathway to suppress inflammation and reduce target organ damage. The physiologic relationship between ACE2 and angiotensin II is promising and can be applied owing to the severity of the situation emerged. Also, drugs inhibiting the activity of TMPRSS2 can be viewed as potential treatment option for COVID-19, as camostat mesilate (TMPRSS2 inhibitor) showed promising results against SARS-CoV-2 infection in mice model (Uno 2020).

3.2 Protease Inhibitor

Chinese National Health Commission recommended Aluvia/Kaletra, as therapy for patients affected with COVID-19 after following the success report of this combination therapy in South Korea. Through this combination therapy, lopinavir and ritonavir (LPVr) which are approved as potential HIV protease inhibitors are administered into the body (Singh and Chhikara 2014). Ritonavir inhibits cytochrome P450 3A metabolism and enhances the half-life of lopinavir. Proteases (M^{pro} or 3CL^{pro}) are involved in viral replication and peptide translation to yield functional viral protein (Yao et al. 2020a). LPVr inhibits the action 3CL^{pro} and blocks the viral replication process. Lim et al. (2020) have conducted a study in Korea by administering the combined therapy to COVID-19 patient and observed a significant decrease in loads of β-coronavirus. The efficacy of protease inhibitors on

COVID-19 was checked in very few patients, and detailed analysis in large groups needs to be done so as to recommend this formulation for treating COVID-19.

3.3   Replication Inhibitors

SARS-CoV-2 is a positive strand RNA virus, which uses host cell machinery for replication through multi-subunit transcription complex consisting of nonstructural proteins (nsp). Among these, nsp12 (RNA polymerase), nsp13 (helicase), nsp14 (fidelity control), and nsp16 (mRNA capping) are rare and unique domains involved in the formation of "nsp interactome," supported by a couple of important and smaller subunits nsp7–nsp10 as cofactors (Snijder et al. 2016). The core subunit of this interactome is nsp12 acting as catalytic subunit of RNA-dependent RNA polymerase (RdRp) which permits the viral genome to transcribe into new RNA templates thereby increasing the infection inside host body (Kirchdoerfer and Ward, 2019). In this connection, nucleotide analogues such as remdesivir (capable of inhibiting polymerase) can be considered as potential target for treating SARS-CoV-2 as antiviral drug.

Yin et al. (2020), through cryo-EM studies, reported that the ability of remdesivir in its triphosphate form at 1 mM has successfully inhibited the RdRp polymerization activity, wherein individually even at 5 mM concentration as prodrug it did not have the sufficient inhibitory effect. This is because the triphosphate form of remdesivir contends with its counterpart, and better results were observed with incorporation of two nucleotide analogues at i and i + 3 position. This approach not only inhibited RNA synthesis but also led to the RNA chain termination (Gordon et al. 2020). Lu et al. (2020a) and Sheahan et al. (2020) after conducting separate in vitro cell based assays suggested the efficacy of galidesivir, ribavirin, favipiravir, and EIDD-2801 in inhibiting the replication of SARS-CoV-2.

Favipiravir (6-fluoro-3-hydroxy-2-pyrazinecarboxamide) is also considered as one of the promising drugs, because, intracellular phosphoribosylation allows the drug to become active and straight away to inhibit RNA polymerization (Furuta et al. 2017). Sheahan et al. (2020) emphasized the 3–10 times better potentiality of EIDD-2801 compared to remdesivir in blocking viral replication due to the structural luxury of possessing extra hydrogen bond with the side chain of K545 and cytidine bases from the template strand. Besides RdRp has several conservative sequences and motifs that share structural similarity with DNA-dependent DNA polymerase, DdRP, and reverse transcriptase (RT). Hence, researchers are also working on DNA synthesis inhibitors such as lamivudine and tenofovir disoproxil, as alternative drugs for inhibiting virus replication through in silico studies (Chang et al. 2020b).

3.4   Heterocyclic Antivirals

Heterocyclic compounds are found to play a significant role in medical chemistry due to their biological activity against protozoa, bacteria, fungi, and especially virus.

Many heterocyclic compounds such as chloroquine, umifenovir (Arbidol), galidesivir, garunavir, oseltamivir, etc. have been used in treating viral infection and pertaining the current situation wherein lack of specific medication led the researchers into experimentation on trial and error basis (Li et al. 2020a; Preethi et al. 2008). Based on the report released on February 2020, among the tested drugs, chloroquine was successful in inhibiting the severity of acute respiratory syndrome coronavirus 2 (SARS-CoV-2) at in vitro conditions.

Chloroquine is a 4-aminoquinoline with antimalarial properties, antiviral, anti-inflammatory, and potential chemosensitization and radiosensitization activities (Gupta et al. 2018; Sharma et al. 2019). At in vitro conditions, it was successful in either interruption of viral replication or inhibition of post-translational modifications/glycosylation of newly synthesised proteins including coronavirus-2, HIV, and chikungunya virus (Savarino et al. 2003). The recommended dose of chloroquine for COVID-19 patient is 1000 mg/ day for a maximum of 10 days the same as prescribed for malarial patients (Riou et al. 1988). Distribution rate of chloroquine is very high and interferes with the metabolic activities of human body for more than 30–60 days (Ducharme and Farinotti 1996). The lethal concentration of drug is 5 g in an adult, and the increase in concentration can cause retinopathy, psychiatric symptoms, and immunosuppression with contraindications in several conditions including pregnancy (Touret and Lamballerie 2020). In contrast, hydroxychloroquine is considered as less toxic with improved efficacy and comparatively at a lower dosage of 600 mg/day can reduce the viral impact (Colson et al. 2020). However, accurate clinical data in humans to assess its effectiveness is required before establishing it as an authenticated treatment.

Based on the preliminary data reported by Wang et al. (2020b), chloroquine in combination with remdesivir shows positive effect of substantially controlling morbidity and mortality due to COVID-19. Further, Gao et.al (2020) conducted multicentric trials related to treating COVID-19 associated with pneumonia and recommended chloroquine as a potential candidate.

3.5 Biological Therapy

Biological therapy includes the usage of direct living organisms, derivatives of living organisms, or biosimilars to treat diseases. This approach is also called "immunotherapy" or "targeted therapy," because it stimulates the immune system of host body via vaccination or transfusion of convalescent plasma.

Vaccines for preventing SARS-CoV-2 are not yet available, but live-attenuated vaccines and protein cage nanoparticles designed for SARS CoV can be evaluated for SARS-CoV-2 owing to their phylogenetic relatedness. In addition to protein cage nanoparticles, rhesus θ-defensin 1, an innate immunomodulator with high anti-SARS-CoV efficiency was proposed to be used as vaccine for SARS-CoV-2 (Wiley et al. 2009). Du et al. (2009) used the similar concept and identified T- and B-cell epitopes specific to spike and nucleocapsid protein identical to SARS-CoV-2 proteins and evaluated for vaccine production. Presently, various technology platforms evaluated for development of vaccines for COVID-19 include DNA, RNA, virus-like particles, viral vectors, inactivated/live attenuated virus, recombinant proteins, and peptides.

The genetic sequence of SARS-CoV-2 was published in January 2020 which triggered the scientific community toward vaccine development against COVID-19 (Le et al. 2020). Upon considering the available data till 8 April 2020, 115 potential vaccine developers are working on the vaccine for COVID-19 among which 78 are confirmed as active and 73 of them are in preclinical stage. The important vaccine candidates include mRNA-1273, Ad5-nCoV, INO-4800, LV-SMENP-DC, and pathogen-specific aAPC (artificial antigen-presenting cell), and their vaccine characteristics are LNP (lipid nanoparticle)-encapsulated mRNA vaccine encoding S protein, adenovirus type 5 vector that expresses S protein, DNA plasmid encoding S protein, DCs (dendritic cell) modified with lentiviral vector expressing synthetic minigene, and aAPCs modified with lentiviral vector expressing synthetic minigene. Few platforms require adjuvants to enhance their immunogenicity, and their administration even at low doses will serve the purpose without comprising the people's safety.

Unavailability of specific antiviral drugs and vaccines as most of them being under clinical trials is an immense need for alternate strategies such as convalescent plasma (CP) therapy to face the present situation. CP therapy is well known from two decades and was successful in treating SARS, MERS, and H1N1. As these viruses share similar virological and clinical characteristics with COVID-19, the researches strongly believe that CP therapy can become a promising rescue for COVID-19-infected patients (Chen et al. 2020a). Duan et al. (2020) conducted a pilot study to explore the feasibility of CP therapy in 10 patients affected severely with COVID-19. Convalescent plasma (200 mL) with high titer of neutralizing antibody (>1:640) was transfused to patients at a median of 16.5 days (Duan et al. 2020). They emphasized that CP therapy besides being safe increases oxygen saturation and lymphocyte count and decreases viral load, C-reactive protein, and lung lesions. Zhang and his colleagues transfused 200–2400 mL of convalescent plasma into four COVID-19-effected people after 11 to 18 days of disease identification. They concluded that all the four patients recovered from COVID-19 in a span of 1 week to 1 month (Zhang et al. 2020a).

Presently, several countries such as the USA, China, South Korea, and the UK are experimenting with CP therapy, and India is also following their footsteps. Kerala is the first state in India to initiate the protocol for CP therapy but waiting for the approval from Drug Controller General of India (DCGI) to relax the norms related to blood donation. Another plasma therapy trial was done on a 49-year-old COVID-19 patient at Max Healthcare in Saket, New Delhi. The patient who was on ventilator recovered and was discharged from the hospital as a result of this therapy. However, Indian Council of Medical Research (ICMR) stated clearly that no therapy including plasma therapy was approved for COVID-19 (Table 1).

Table 1 Pharmacological therapies for COVID 19

Drug therapy	Classification	Mode of action	Safety concerns	References
Chloroquine hydroxychloroquine	Antimalarial	Inhibits the activity of DNA and RNA polymerases Interferes with post-translation of viral protein and synthesis of virus Interferes with viral particle from binding to ACE2 cellular receptor Mediates anti-inflammatory response	Risk of heart arrhythmias and retinal damage Caution in diabetics and patients with Glucose-6-phosphate dehydrogenase (G6PD) deficiency	Wang et al. (2020b), Yao et al. (2020b)
Lopinavir Ritonavir	HIV protease inhibitor	Binds to M^{pro} and suppresses its activity and prevents coronavirus replication	Risk of cardiac arrhythmias Patients with hepatitis are at risk	Yao et al. (2020a)
Remdesivir	Nucleoside analogue	The coronavirus susceptibility toward remdesivir is mediated by viral polymerase and exoribonuclease (ExoN) Adenosine analogue (GS-441524) acts as an inhibitor of RNA-dependent RNA polymerase, avoids proofreading by ExoN and prevents RNA synthesis	Hepatotoxicity is an identified risk Causes transaminitis	Brown et al. (2019), Ko et al. (2020)
Favipiravir	RNA-dependent RNA polymerase inhibitor	Within the cell, favipiravir is converted into favipiravir-RTP, which is recognized as substrate by viral RNA polymerase. The inhibition of RNA polymerase activity prevents viral RNA elongation	Favipiravir has potential to cause teratogenicity and embryotoxicity in humans	Dong et al. (2020)

Azithromycin	Macrolide type antibacterial	Azithromycin downregulates the cytokine production which trigger lung inflammation in COVID 19 patients It accelerate phagocytosis ability of macrophages Along with hydroxychloroquine it inhibit the replication of corona virus	Risk of heart arrhythmias, jaundice, and allergic reactions	Zarogoulidis et al. (2012)
Tocilizumab Sarilumab Siltuximab	Anti-IL-6 drugs	These drugs binds with both soluble and membrane-bound IL-6 receptors and inhibits IL-6-mediated signaling It activates T cell and macrophages, enhances immunoglobulin secretion and hepatic acute-phase protein synthesis	Patients with rheumatoid arthritis are at high risk	Zhang et al. (2020b)
Leronlimab	Human monoclonal antibody	It enhances immune response while mitigating cytokine storm	No major safety concerns	Miao et al. (2020)
CP therapy	Plasma from COVID-19 recovered patient contain antibodies against SARS-CoV-2	Antibody binds to the pathogen and neutralizes its infectivity directly, or through complement activation, phagocytosis and antibody-dependent cellular cytotoxicity This therapy confers immediate immunity to susceptible individuals	Logistical challenges No major safety concerns	Keith et al. (2020)
Corticosteroids and inhaled pulmonary vasodilators	Aerosolized drugs	This treatment is considered for patients with acute respiratory distress syndrome	Increase the risk of pneumonia and osteoporotic fractures	Suissa et al. (2013)

(continued)

Table 1 (continued)

Drug therapy	Classification	Mode of action	Safety concerns	References
Anticoagulation	Clot busters	Venous thromboembolism prophylaxis with low molecular weight heparin and thrombolytic therapy such as tissue plasminogen activator is recommended for COVID-19 patients	Patient with acute medical illness are t high risk	Tang et al. (2020)
Non-steroidal anti-inflammatory drugs (NSAIDS)	Analgesic	Indomethacin (at concentration of 1 mg/kg) exhibits potent antiviral activity against human SARS-CoV by inhibiting virus replication	It increases risk of heart stroke if used for long term at high dose	Russell et al. (2020)
Bronchodilators	Beta 2-agonists and anticholinergics	They stimulate muscle beta-2 receptors that results in relaxation and widens the airways	Caution in patients with hyperthyroidism, cardiovascular disease, arrhythmia, hypertension, and diabetes	Suissa et al. (2013)

3.6 Herbal Therapeutics

The use of herbals for treating people with various diseases has a long history (Shaheena et al. 2019). The effectiveness of these herbals depends on the bioactive phytochemicals which they possess (Naidu et al. 2016; Sravani et al. 2015; Raju et al. 2019; Kumar et al. 2019a). These bioactive chemicals include flavonoids, tannins, alkaloids, essential oils, and other secondary metabolites (Dirisala et al. 2017; Kumar et al. 2019a, b, 2020). In the past seven decade, several reports have come out on antiviral efficacy of medicinal plant extracts. In 1995, McCutcheon et al. reported antiviral activity of medicinal plants against number of viruses including coronaviruses (McCutcheon et al. 1995) (Fig. 1).

Numerous literatures on traditional Chinese medicine (TCM) treatments for SARS-CoV infection are published after the SARS epidemic in China (Yang et al. 2020b). Herbal extracts and bioactive compounds from many Chinese herbs have been documented to exhibit anti-SARS-CoV activity including extracts from *Artemisia annua*, *Lycoris radiata*, *Pyrrosia lingua*, *Houttuynia cordata*, and *Lindera aggregate* (Li et al. 2005) and myricetin, scutellarein, and other phenolic compounds from *Isatis indigotica* and *Torreya nucifera* (Lin et al. 2005; Ryu et al. 2010; Yu et al. 2012). A formulation of Indian system of traditional medicine Kabasurakudineer (KSK) which comprises of 15 herbs was found to be effective against swine flu (Uthamarayan 2006; Mudaliar 2013; Kaba Suram 2020). A number of hydrolysable tannins are explored to have anti-COVID-19 property through an in silico approach (Khalifa et al. 2020).

Ryu et al. (2010) reported that *T. nucifera* plant's biflavonoids have anti-SARS-CoV activity as it showed inhibition property against SARS-CoV chymotrypsin-like protease (3CL^{pro}) (Ryu et al. 2010). *Psoralea corylifolia* seeds were found to have anti-SRAS-CoV property as their ethanolic extract showed inhibiting property against SARS-CoV-papain-like protease (PL^{pro}) (Kim et al. 2014). Due to epidemiological, genomics, and pathogenesis homology between SARS-CoV and SARS-CoV-2 (Zhou et al. 2020b; Wu et al. 2020), the herbals with anti-SARS-CoV activity can potentially work against SARS-CoV-2 infection. Recently, applying molecular docking technique, it has been reported that garlic essential oil has resistance property against SARS-CoV-2 (Thuy et al. 2020). Many other bioactive phytochemicals have also been explored through docking and other bioinformatics approach and appeared to have potential as COVID-19- main protease (M^{pro}) inhibitors; these compounds are luteolin-7-glucoside, curcumin, oleuropein, demethoxycurcumin, rhein, catechin, apigenin-7-glucoside, epicatechin-gallate, withaferin-A, withanolide-D, and Aloe-emodin (Khaerunnisa et al. 2020; Chandel et al. 2020).

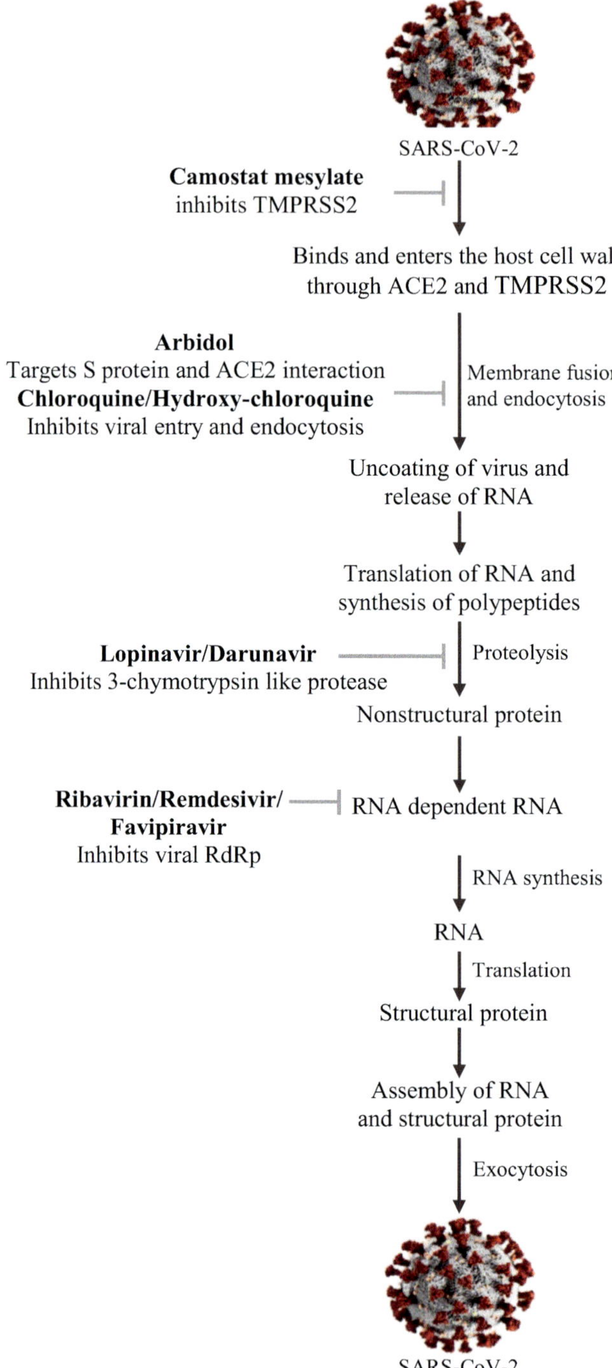

Fig. 1 Simplified diagram showing life cycle of SARS-CoV-2 and potential drug targets

4   Conclusion

The COVID-19 pandemic created a worldwide health emergency. For management of this pandemic, timely accurate detection, isolation, and prompt treatment as well as contact tracing are very important. As the disease symptoms resemble with other respiratory infections, differential diagnosis comes into play. Molecular diagnostic can be adopted as first line of differential testing for COVID-19, while chest CT can be considered for rapid testing of the disease. Now, exposure history is not of much relevance for COVID-19 diagnosis in this pandemic situation but still could be considered as supportive diagnosis. Further, as the information on SARS-CoV-2 are enriching, many potential therapeutics are being explored particularly plasma treatment. Exploring newer methods would certainly pave way for identification of therapeutics. Besides, the promising therapeutics mentioned in the chapter are going through the clinical trials; upon observations of positive results, these therapeutics can be approved and applied with complete effective treatment.

Acknowledgments   The authors are grateful to Director, ICAR-IISS, Mau; Management, VFSTR, Guntur, and Dean, COA, Pasighat and VC, CAU, Imphal for extending their kind support in writing the timely chapter.

References

Ai T, Yang Z, Hou H, Zhan C, Chen C, Lv W, Tao Q, Sun Z, Xia L (2020) Correlation of chest CT and RT-PCR testing in coronavirus disease 2019 (COVID-19) in China: a report of 1014 cases. Radiology. https://doi.org/10.1148/radiol.2020200642

Brown AJ, Won JJ, Graham RL, Dinnon KH III, Sims AC, Feng JY, Cihlar T, Denison MR, Baric RS, Sheahan TP (2019) Broad spectrum antiviral remdesivir inhibits human endemic and zoonotic deltacoronaviruses with a highly divergent RNA dependent RNA polymerase. Antivir Res 169:104541

Centers for Disease Control and Prevention (2020a) Coronavirus Disease 2019. https://www.cdc.gov/coronavirus/2019-ncov/prevent-getting-sick/prevention.html

Centers for Disease Control and Prevention (2020b) Coronavirus disease 2019 (COVID-19). Older adults. https://www.cdc.gov/coronavirus/2019-ncov/need-extra-precautions/older-adults.html

Chan JF, Kok KH, Zhu Z, Chu H, To KK, Yuan S, Yuen KY (2020a) Genomic characterization of the 2019 novel human-pathogenic coronavirus isolated from a patient with atypical pneumonia after visiting Wuhan. Emerg Microbes Infect 9:221–236

Chan JF, Yuan S, Kok KH, To KKW, Chu H, Yang J, Xing F, Liu J, Yip CCY, Poon RWS, Tsoi HW, Lo SKF, Chan KH, Poon VKM, Chan WM, Ip JD, Cai JP, Cheng VCC, Chen H, Hui CKM, Yuen KY (2020b) A familial cluster of pneumonia associated with the 2019 novel coronavirus indicating person-to-person transmission: a study of a family cluster. Lancet 395:514–523

Chandel V, Raj S, Rathi B, Kumar D (2020) In silico identification of potent FDA approved drugs against coronavirus COVID-19 main protease: a drug repurposing approach. Chem Biol Lett 7:166–175

Chang D, Lin M, Wei L, Xie L, Zhu G, Dela-Cruz CS, Sharma L (2020a) Epidemiologic and clinical characteristics of novel coronavirus infections involving 13 patients outside Wuhan, China. JAMA 323:1092–1093

Chang YC, Tung YA, Lee KH, Chen TF, Hsiao YC, Chang HC, Hsieh TT, Su CH, Wang SS, Yu JY, Shih SS (2020b) Potential therapeutic agents for COVID-19 based on the analysis of protease and RNA polymerase docking

Chawla LS, Chen S, Bellomo R, Tidmarsh GF (2018) Angiotensin converting enzyme defects in shock: implications for future therapy. Crit Care 22:274

Chen L, Xiong J, Bao L, Shi Y (2020a) Convalescent plasma as a potential therapy for COVID-19. Lancet Infect Dis 20:398–400

Chen N, Zhou M, Dong X, Qu J, Gong F, Han Y, Qiu Y, Wang J, Liu Y, Wei Y, Yu T (2020b) Epidemiological and clinical characteristics of 99 cases of 2019 novel coronavirus pneumonia in Wuhan, China: a descriptive study. Lancet 395:507–513

Chhikara BS, Rathi B, Singh J, Poonam FNU (2020) Corona virus SARS-CoV-2 disease COVID-19: infection, prevention and clinical advances of the prospective chemical drug therapeutics. Chem Biol Lett 7:63–72

Colson P, Rolain JM, Lagier JC, Brouqui P, Raoult D (2020) Chloroquine and hydroxychloroquine as available weapons to fight COVID-19. Int J Antimicrob Agents 55:105932

Coronavirus disease (COVID-19) outbreak situation, 2020. May 2. Retrieved from https://www.who.int/emergencies/diseases/novel-coronavirus-2019

Dirisala VR, Nair RR, Srirama K, Reddy PN, Rao KS, Kumar NSS, Parvatam G (2017) Recombinant pharmaceutical protein production in plants: unraveling the therapeutic potential of molecular pharming. Acta Physiol Plant 39:18

Dong L, Hu S, Gao J (2020) Discovering drugs to treat coronavirus disease 2019 (COVID-19). Drug Discov Ther 14:58–60

Du L, He Y, Zhou Y, Liu S, Zheng BJ, Jiang S (2009) The spike protein of SARS-CoV—a target for vaccine and therapeutic development. Nat Rev Microbiol 7:226–236

Duan K, Liu B, Li C, Zhang H, Yu T, Qu J, Zhou M, Chen L, Meng S, Hu Y, Peng C (2020) Effectiveness of convalescent plasma therapy in severe COVID-19 patients. Proc Natl Acad Sci 117:9490–9496

Ducharme J, Farinotti R (1996) Clinical pharmacokinetics and metabolism of chloroquine. Clin Pharmacokinet 31:257–274

European Centre for Disease Prevention and Control (2020) Outbreak of novel coronavirus disease 2019 (COVID-19): increased transmission globally–fifth update, 2 Mar 2020

Fernandes T, Hashimoto NY, Magalhães FC, Fernandes FB, Casarini DE, Carmona AK, Krieger JE, Phillips MI, Oliveira EM (2011) Aerobic exercise training–induced left ventricular hypertrophy involves regulatory MicroRNAs, decreased angiotensin-converting enzyme-angiotensin II, and synergistic regulation of angiotensin-converting enzyme 2-angiotensin (1-7). Hypertension 58:182–189

Ferrario CM, Jessup J, Chappell MC, Averill DB, Brosnihan KB, Tallant EA, Diz DI, Gallagher PE (2005) Effect of angiotensin-converting enzyme inhibition and angiotensin II receptor blockers on cardiac angiotensin-converting enzyme 2. Circulation 111:2605–2610

Fu Y, Cheng Y, Wu Y (2020) Understanding SARS-CoV-2-mediated inflammatory responses: from mechanisms to potential therapeutic tools. Virol Sin 35:266–271

Furuta Y, Komeno T, Nakamura T (2017) Favipiravir (T-705), a broad spectrum inhibitor of viral RNA polymerase. Proc Jpn Acad Ser B 93:449–463

Gao J, Tian Z, Yang X (2020) Breakthrough: chloroquine phosphate has shown apparent efficacy in treatment of COVID-19 associated pneumonia in clinical studies. Biosci Trends 14:72–73

Gordon CJ, Tchesnokov EP, Feng JY, Porter DP, Götte M (2020) The antiviral compound remdesivir potently inhibits RNA-dependent RNA polymerase from Middle East respiratory syndrome coronavirus. J Biol Chem 295:4773–4779

Guan WJ, Ni ZY, Hu Y, Liang WH, Ou CQ, He JX, Liu L, Shan H, Lei CL, Hui DS, Du B (2020) Clinical characteristics of coronavirus disease 2019 in China. N Engl J Med 58:711–712

Guarner J (2020) Three emerging coronaviruses in two decades-the story of SARS, MERS, and now COVID-19. Am J Clin Pathol 153:420–421

Gupta Y, Gupta N, Singh S, Wu L, Chhikara BS, Rawat M, Rathi B (2018) Multistage inhibitors of the malaria parasite: emerging hope for chemoprotection and malaria eradication. Med Res Rev 38:1511–1535

Huang C, Wang Y, Li X, Ren L, Zhao J, Hu Y, Zhang L, Fan G, Xu J, Gu X, Cheng Z (2020) Clinical features of patients infected with 2019 novel coronavirus in Wuhan, China. Lancet 395:497–506

Ishiyama Y, Gallagher PE, Averill DB, Tallant EA, Brosnihan KB, Ferrario CM (2004) Upregulation of angiotensin-converting enzyme 2 after myocardial infarction by blockade of angiotensin II receptors. Hypertension 43:970–976

Jiang S, Du L, Shi Z (2020) An emerging coronavirus causing pneumonia outbreak in Wuhan, China: calling for developing therapeutic and prophylactic strategies. Emerg Microbes Infect 9:275–277

Jordan RE, Adab P, Cheng KK (2020) Covid-19: risk factors for severe disease and death. BMJ 368:m1198

Kaba Suram (Swine Flu) (2020). https://www.nhp.gov.in/swine-flu_mtl. Accessed on 3 May 2020

Keith P, Day M, Perkins L, Moyer L, Hewitt K, Wells A (2020) A novel treatment approach to the novel coronavirus: an argument for the use of therapeutic plasma exchange for fulminant COVID-19

Khaerunnisa S, Kurniawan H, Awaluddin R, Suhartati S, Soetjipto S (2020) Potential inhibitor of COVID-19 Main protease (Mpro) from several medicinal plant compounds by molecular docking study. Preprints 2020030226. https://doi.org/10.20944/preprints202003.0226.v1

Khalifa I, Zhu W, Nafie MS, Dutta K, Li C (2020) Anti-COVID-19 effects of ten structurally different hydrolysable tannins through binding with the catalytic-closed sites of COVID-19 main protease: an *In-silico* approach. Preprints 2020030277. https://doi.org/10.20944/preprints202003.0277.v1

Kim DW, Seo KH, Curtis-Long MJ, Oh KY, Oh JW, Cho JK, Lee KH, Park KH (2014) Phenolic phytochemical displaying SARS-CoV papain-like protease inhibition from the seeds of *Psoralea corylifolia*. J Enzyme Inhib Med Chem 29:59–63

Kirchdoerfer RN, Ward AB (2019) Structure of the SARS-CoV nsp12 polymerase bound to nsp7 and nsp8 co-factors. Nat Commun 10:1–9

Ko WC, Rolain JM, Lee NY, Chen PL, Huang CT, Lee PI, Hsueh PR (2020) Arguments in favour of remdesivir for treating SARS-CoV-2 infections. Int J Antimicrob Agents 55(4):105933

Kumar A, Agarwal DK, Kumar S, Reddy YM, Chintagunta AD, Saritha KV, Pal G, Kumar SJ (2019a) Nutraceuticals derived from seed storage proteins: implications for health wellness. Biocatal Agric Biotechnol 17:710–719

Kumar A, Ramesh KV, SRP C, Sripathy KV, Agarwal DK, Pal G, Kuchlan MK, Singh RK, Ratnaprabha, Kumar SPJ (2019b) Bio-prospecting nutraceuticals from selected soybean skins and cotyledons. Indian J Agric Sci 89(12):2064–2068

Kumar SPJ, Kumar A, Ramesh KV, SRP C, Agarwal DK, Pal G, Kuchlan MK, Singh RK (2020) Wall bound phenolics and total antioxidants in stored seeds of soybean genotypes. Indian J Agric Sci 90(01):118–122

Lai CC, Shih TP, Ko WC, Tang HJ, Hsueh PR (2020) Severe acute respiratory syndrome coronavirus 2 (SARS-CoV-2) and coronavirus disease 2019 (COVID 19): the epidemic and the challenges. Int J Antimicrob Agents 55:105924

Le TT, Andreadakis Z, Kumar A, Román RG, Tollefsen S, Saville M, Mayhew S (2020) The COVID-19 vaccine development landscape. Nat Rev Drug Discov 19:305–306

Lei P, Fan B, Wang P (2020) Differential diagnosis for coronavirus disease (COVID-19): beyond radiologic features. Am J Roentgenol 215:W1

Li Y, Xia L (2020) Coronavirus disease 2019 (COVID-19): role of chest CT in diagnosis and management. Am J Roentgenol 214:1–7

Li W, Moore MJ, Vasilieva N, Sui J, Wong SK, Berne MA, Somasundaran M, Sullivan JL, Luzuriaga K, Greenough TC, Choe H (2003) Angiotensin-converting enzyme 2 is a functional receptor for the SARS coronavirus. Nature 426:450–454

Li SY, Chen C, Zhang HQ, Guo HY, Wang H, Wang L, Zhang X, Hua S, Yu J, Xiao P, Li R, Tan X (2005) Identification of natural compounds with antiviral activities against SARS-associated coronavirus. Antivir Res 67:18–23

Li H, Wang YM, Xu JY, Cao B (2020a) Potential antiviral therapeutics for 2019 novel coronavirus. Zhonghua jie he he hu xi za zhi= Zhonghua jiehe he huxi zazhi=. Chin J Tuberc Respir Dis 43: E002–E002

Li Q, Guan X, Wu P, Wang X, Zhou L, Tong Y, Ren R, Leung KSM, Lau EHY, Wong JY, Xing X, Xiang N, Wu Y, Li C, Chen Q, Li D, Liu T, Zhao J, Liu M, Tu W, Chen C, Jin L, Yang R, Wang Q, Zhou S, Wang R, Liu H, Luo Y, Liu Y, Shao G, Li H, Tao Z, Yang Y, Deng Z, Liu B, Ma Z, Zhang Y, Shi G, Lam TTY, Wu JT, Gao GF, Cowling BJ, Yang B, Leung GM, Feng Z (2020b) Early transmission dynamics in Wuhan, China, of novel coronavirus-infected pneumonia. N Engl J Med 382:1199–1207

Lim J, Jeon S, Shin HY, Kim MJ, Seong YM, Lee WJ, Choe KW, Kang YM, Lee B, Park SJ (2020) Case of the index patient who caused tertiary transmission of coronavirus disease 2019 in Korea: the application of lopinavir/ritonavir for the treatment of COVID-19 pneumonia monitored by quantitative RT-PCR. J Korean Med Sci 35:e79–e79

Lin CW, Tsai FJ, Tsai CH, Lai CC, Wan L, Ho TY, Hsieh CC, Chao PDL (2005) Anti-SARS coronavirus 3C-like protease effects of *Isatis indigotica* root and plant-derived phenolic compounds. Antivir Res 68:36–42

Liu J, Liao X, Qian S, Yuan J, Wang F, Liu Y, Wang Z, Wang FS, Liu L, Zhang Z (2020) Community transmission of severe acute respiratory syndrome coronavirus 2, Shenzhen, China, 2020. Emerg Infect Dis. https://doi.org/10.3201/eid2606.200239

Lu CC, Chen MY, Chang YL (2020a) Potential therapeutic agents against COVID-19: what we know so far. J Chin Med Assoc 83:534–536

Lu H, Stratton CW, Tang Y (2020b) Outbreak of pneumonia of unknown etiology in Wuhan China: the mystery and the miracle. J Med Virol 92:401–402

Mahapatra S, Chandra P (2020) Clinically practiced and commercially viable nanobio engineered analytical methods for COVID-19 diagnosis. Biosens Bioelectron 165:112361

McCutcheon AR, Roberts TE, Gibbons E, Ellis SM, Babiuk LA, Hancock RE, Towers GH (1995) Antiviral screening of British Columbian medicinal plants. J Ethnopharmacol 49:101–110

Miao M, De Clercq E, Li G (2020) Clinical significance of chemokine receptor antagonists. Expert Opin Drug Metab Toxicol 16:11–30

Mudaliar MKS (2013) Siddha Materia Medica, Directorate of Indian Medicine & Homeopathy, Chennai-106

Naidu NK, Ramu VV, Kumar NS (2016) Anti-inflammatory and anti-helminthic activity of ethanolic extract of *Azadirachta indica* leaves. Int J Green Pharm 10:S1–S4

Pan Y, Zhang D, Yang P, Poon LLM, Wang Q (2020) Viral load of SARS-CoV-2 in clinical samples. Lancet Infect Dis 20:411–412

Paules CI, Marston HD, Fauci AS (2020) Coronavirus infections- more than just the common cold. JAMA:E1–E2

Peeri NC, Shrestha N, Rahman MS, Zaki R, Tan Z, Bibi S, Baghbanzadeh M, Aghamohammadi N, Zhang W, Haque U (2020) The SARS, MERS and novel coronavirus (COVID-19) epidemics, the newest and biggest global health threats: what lessons have we learned? Int J Epidemiol 49:1–10

Preethi D, Anand PS, Kumar SPJ, Josthna P, Naidu CV (2008) In vitro plant regeneration of *Stevia rebaudiana*. J Trop Med Plants 9(1):71–76

Raju NV, Sukumar K, Reddy GB, Pankaj PK, Muralitharan G, Annapareddy S (2019) In-vitro studies on antitumour and antimicrobial activities of Methanolic kernel extract of Mangifera indica L. cultivar Banganapalli. Biomed Pharmacol J 12:357–362

Ren LL, Wang YM, Wu ZQ, Xiang ZC, Guo L, Xu T, Xu T, Jiang YZ, Xiong Y, Li YJ, Li XW, Li H, Fan GH, Gu XY, Xiao Y, Gao H, Xu JY, Yang F, Wang XM, Wu C, Chen L, Liu YW, Liu B, Yang J, Wang XR, Dong J, Li L, Huang CL, Zhao JP, Hu Y, Cheng ZS, Liu LL, Qian

ZH, Qin C, Jin Q, Cao B, Wang JW (2020) Identification of a novel coronavirus causing severe pneumonia in human: a descriptive study. Chin Med J 133:1015–1024

Riou B, Barriot P, Rimailho A, Baud FJ (1988) Treatment of severe chloroquine poisoning. N Engl J Med 318:1–6

Rothe C, Schunk M, Sothmann P, Bretzel G, Froeschl G, Wallrauch C, Zimmer T, Thiel V, Janke C (2020) Transmission of 2019-nCoV infection from an asymptomatic contact in Germany. N Engl J Med 382:970–971

Russell B, Moss C, Rigg A, Van Hemelrijck M (2020) COVID-19 and treatment with NSAIDs and corticosteroids: should we be limiting their use in the clinical setting? Ecancermedicalscience 14:1023

Ryu YB, Jeong HJ, Kim JH, Kim YM, Park JY, Kim D, Naguyen TTH, Park SJ, Chang JS, Park KH, Rho MC, Lee WS (2010) Biflavonoids from *Torreya nucifera* displaying SARS-CoV 3CL (pro) inhibition. Bioorg Med Chem Lett 18:7940–7947

Savarino A, Boelaert JR, Cassone A, Majori G, Cauda R (2003) Effects of chloroquine on viral infections: an old drug against today's diseases. Lancet Infect Dis 3:722–727

Shaheena S, Chintagunta AD, Dirisala VR, Kumar NS (2019) Extraction of bioactive compounds from *Psidium guajava* and their application in dentistry. AMB Express 9:208

Sharma N, Verma A, Poonam FNU, Kempaiah P, Rathi B (2019) Chemical libraries targeting liver stage malarial infection. Chem Biol Lett 6:14–22

Sheahan TP, Sims AC, Zhou S, Graham RL, Pruijssers AJ, Agostini ML, Leist SR, Schäfer A, Dinnon KH, Stevens LJ, Chappell JD (2020) An orally bioavailable broad-spectrum antiviral inhibits SARS-CoV-2 in human airway epithelial cell cultures and multiple coronaviruses in mice. Sci Transl Med 12(541):1–15, eabb5883

Singh J, Chhikara BS (2014) Comparative global epidemiology of HIV infections and status of current progress in treatment. Chem Biol Lett 1:14–32

Singhal T (2020) A review of coronavirus Disease-2019 (COVID-19). Indian J Pediatr 87:281–286

Snijder EJ, Decroly E, Ziebuhr J (2016) The nonstructural proteins directing coronavirus RNA synthesis and processing. In: Advances in virus research, vol 96. Academic, New York, pp 59–126

Sohrabi C, Alsafi Z, O'Neill N, Khan M, Kerwan A, Al-Jabir A, Iosifidis C, Agha R (2020) World Health Organization declares global emergency: a review of the 2019 novel coronavirus (COVID-19). Int J Surg 76:71–76

Sravani D, Aarathi K, Kumar NS, Krupanidhi S, Ramu DV, Venkateswarlu TC (2015) In vitro anti-inflammatory activity of Mangifera indica and Manilkara zapota leaf extract. Res J Pharm Technol 8:1477–1480

Suissa S, Patenaude V, Lapi F, Ernst P (2013) Inhaled corticosteroids in COPD and the risk of serious pneumonia. Thorax 68:1029–1036

Tahamtana A, Ardebili A (2020) Real-time RT-PCR in COVID-19 detection: issues affecting the results. Expert Rev Mol Diagn 20:453–454

Tang N, Bai H, Chen X, Gong J, Li D, Sun Z (2020) Anticoagulant treatment is associated with decreased mortality in severe coronavirus disease 2019 patients with coagulopathy. J Thromb Haemost 18:1094–1099

Thuy BTP, My TTA, Hai NTT, Hieu LT, Hoa TT, Loan HTP, Triet NT, Anh TTV, Quy PT, Tat PV, Hue NV, Quang DT, Trung NT, Tung VT, Huynh LK, Nhung NTA (2020) Investigation into SARS-CoV-2 resistance of compounds in garlic essential oil. ACS Omega 5:8312–8320

Touret F, de Lamballerie X (2020) Of chloroquine and COVID-19. Antivir Res 177:104762

U.S Food & Drug Administration (2020) Dr. Bright, Biomedical Advanced Research and Development Authority. https://www.fda.gov/media/136534/download

Udugama B, Kadhiresan P, Kozlowski HN, Malekjahani A, Osborne M, Li VYC, Chen H, Mubareka S, Gubbay JB, Chan WCW (2020) Diagnosing COVID-19: the disease and tools for detection. ACS Nano 14:3822–3835

Uno Y (2020) Camostat mesilate therapy for COVID-19. Intern Emerg Med 29:1–2. https://doi.org/10.1007/s11739-020-02345-9. [Epub ahead of print]

Uthamarayan KS (2006) Uthamarayan KS, siddha Maruthuvanga churukkam, Directorate of Indian Medicine & homeopathy, Chennai-106

Wang D, Hu B, Hu C, Zhu F, Liu X, Zhang J, Wang B, Xiang H, Cheng Z, Xiong Y, Zhao Y, Li Y, Wang X, Peng Z (2020a) Clinical characteristics of 138 hospitalized patients with 2019 novel coronavirus-infected pneumonia in Wuhan, China. JAMA 323:1061–1069

Wang M, Cao R, Zhang L, Yang X, Liu J, Xu M, Shi Z, Hu Z, Zhong W, Xiao G (2020b) Remdesivir and chloroquine effectively inhibit the recently emerged novel coronavirus (2019-nCoV) in vitro. Cell Res 30:269–271

WHO Coronavirus Disease (COVID-19) Dashboard (2020) Global situation. https://covid19.who.int/

Wiley JA, Richert LE, Swain SD, Harmsen A, Barnard DL, Randall TD, Jutila M, Douglas T, Broomell C, Young M, Harmsen A (2009) Inducible bronchus-associated lymphoid tissue elicited by a protein cage nanoparticle enhances protection in mice against diverse respiratory viruses. PLoS One 4(9):e7142

Winichakoon P, Chaiwarith R, Liwsrisakun C, Salee P, Goonna A, Limsukon A, Kaewpoowat Q (2020) Negative nasopharyngeal and oropharyngeal swab does not rule out COVID-19. J Clin Microbiol 58:e00297–e00220

Wong JEL, Leo YS, Tan CC (2020) COVID-19 in Singapore-current experience: critical global issues that require attention and action. JAMA 323:1243–1244

World Health Organisation (2020c) Modes of transmission of virus causing COVID-19: implications for IPC precaution recommendations. Scientific brief 29 Mar 2020

World Health Organisation (2020f) Report of the WHO-China joint Mission on coronavirus disease 2019 (COVID-19), 16–24 Febr 2020

World Health Organisation (2020g) Laboratory testing for coronavirus disease 2019 (COVID-19) in Suspected Human Cases, Interim guidance 2 Mar 2020

World Health Organization (2020a) WHO Director-General's Remarks at the Media Briefing on 2019-nCoV on 11 Feb 2020, available at https://www.who.int/dg/speeches/detail/who-director-general-s-remarks-at-the-media-briefing-on-2019-ncov-on-11-february-2020

World Health Organization (2020b) WHO Director-General's opening remarks at the media briefing on COVID-19 – 11 Mar 2020, available at https://www.who.int/dg/speeches/detail/who-director-general-s-opening-remarks-at-the-media-briefing-on-covid-19%2D%2D-11-march-2020

World Health Organization (2020d) Novel coronavirus (2019-nCoV), Situation Report–12, 1 Feb 2020

World Health Organization (2020e) Novel Coronavirus, Advice for the Public. https://www.who.int/emergencies/diseases/novel-coronavirus-2019/advice-for-public

Wu A, Peng Y, Huang B, Ding X, Wang X, Niu P, Meng J, Zhu Z, Zhang Z, Wang J, Sheng J, Quan L, Xia Z, Tan W, Cheng G, Jiang T (2020) Genome composition and divergence of the novel coronavirus (2019-nCoV) originating in China. Cell Host Microbe 27:325–328

Yang J, Zheng Y, Gou X, Pu K, Chen Z, Guo Q, Ji R, Wang H, Wang Y, Zhou Y (2020a) Prevalence of comorbidities and its effects in coronavirus disease 2019 patients: a systematic review and meta-analysis. Int J Infect Dis 94:91–95

Yang Y, Islam MS, Wang J, Li Y, Chen X (2020b) Traditional Chinese medicine in the treatment of patients infected with 2019-new coronavirus (SARS-CoV-2): a review and perspective. Int J Biol Sci 16:1708–1717

Yang Y, Yang M, Shen C, Wang F, Yuan J, Li J, Zhang M, Wang Z, Xing L, Wei J, Peng L, Wong G, Zheng H, Liao M, Feng K, Li J, Yang Q, Zhao J, Zhang Z, Liu L Liu Y (2020c) Evaluating the accuracy of different respiratory specimens in the laboratory diagnosis and monitoring the viral shedding of 2019-nCoV infections. medRxiv, https://doi.org/10.1101/2020.02.11.20021493

Yao TT, Qian JD, Zhu WY, Wang Y, Wang GQ (2020a) A systematic review of lopinavir therapy for SARS coronavirus and MERS coronavirus—a possible reference for coronavirus disease-19 treatment option. J Med Virol 2:556–563

Yao X, Ye F, Zhang M, Cui C, Huang B, Niu P, Liu X, Zhao L, Dong E, Song C, Zhan S (2020b) In vitro antiviral activity and projection of optimized dosing design of hydroxychloroquine for the treatment of severe acute respiratory syndrome coronavirus 2 (SARS-CoV-2). Clin Infect Dis 71:732–739

Ye F, Xu S, Rong Z, Xu R, Liu X, Deng P, Liu H, Xu X (2020) Delivery of infection from asymptomatic carriers of COVID-19 in a familial cluster. Int J Infect Dis 94:133–138

Yin W, Mao C, Luan X, Shen DD, Shen Q, Su H, Wang X, Zhou F, Zhao W, Gao M, Chang S (2020) Structural basis for the inhibition of the RNA-dependent RNA polymerase from SARS-CoV-2 by Remdesivir. bioRxiv, https://doi.org/10.1101/2020.04.08.032763

Yu MS, Lee J, Lee JM, Kim Y, Chin YW, Jee JG, Keum YS, Jeong YJ (2012) Identification of myricetin and scutellarein as novel chemical inhibitors of the SARS coronavirus helicase, nsP13. Bioorg Med Chem Lett 22:4049–4054

Zarogoulidis P, Papanas N, Kioumis I, Chatzaki E, Maltezos E, Zarogoulidis K (2012) Macrolides: from in vitro anti-inflammatory and immunomodulatory properties to clinical practice in respiratory diseases. Eur J Clin Pharmacol 68:479–503

Zhang L, Liu Y (2020) Potential interventions for novel coronavirus in China: a systematic review. J Med Virol 92:479–490

Zhang B, Liu S, Tan T, Huang W, Dong Y, Chen L, Chen Q, Zhang L, Zhong Q, Zhang X, Zou Y (2020a) Treatment with convalescent plasma for critically ill patients with SARS-CoV-2 infection. Chest 158:e9–e13

Zhang C, Wu Z, Li JW, Zhao H, Wang GQ (2020b) The cytokine release syndrome (CRS) of severe COVID-19 and Interleukin-6 receptor (IL-6R) antagonist tocilizumab may be the key to reduce the mortality. Int J Antimicrob Agents 55:105954

Zhou F, Yu T, Du R, Fan G, Liu Y, Liu Z, Xiang J, Wang Y, Song B, Gu X, Guan L, Wei Y, Li H, Wu X, Xu J, Tu S, Zhang Y, Chen H, Cao B (2020a) Clinical course and risk factors for mortality of adult inpatients with COVID-19 in Wuhan, China: a retrospective cohort study. Lancet 395:1054–1062

Zhou P, Yang XL, Wang XG, Hu B, Zhang L, Zhang W, Si HR, Zhu Y, Li B, Huang CL, Chen HD, Chen J, Luo Y, Guo H, Jiang RD, Liu MQ, Chen Y, Shen XR, Wang X, Zheng XS, Zhao K, Chen QJ, Deng F, Liu LL, Yan B, Zhan FX, Wang YY, Xiao GF, Shi ZL (2020b) A pneumonia outbreak associated with a new coronavirus of probable bat origin. Nature 579:270–273

Zhu N, Zhang D, Wang W, Li X, Yang B, Song J, Zhao X, Huang B, Shi W, Lu R, Niu P, Zhan F, Ma X, Wang D, Xu W, Wu G, Gao GF, Tan W (2020) A novel coronavirus from patients with pneumonia in China, 2019. N Engl J Med 382:727–733

Ziaie S, Koucheck M, Miri MM, Salarian S, Shojaei S, Haghighi M, Sistanizad M (2020) Review of therapeutic agents for treatment of COVID-19. J Cell Mol Anesth 5:32–36

Overview of Coronavirus Disease and Imaging-Based Diagnostic Techniques

Archana Ramadoss, Veena Raj, and Mithun Kuniyil Ajith Singh

Abstract

Human coronaviruses are highly contagious pathogens that cause acute respiratory syndrome in humans. Detailed characterization of the pathogen and appropriate detection techniques are critical for rapid clinical decision making in order to treat the infected individuals and also to prevent further spread of the disease. Recent pandemic caused by SARS-CoV2 that causes COVID-19 disease is an example that demonstrates the severity of such a crisis. Coronavirus genome consists of numerous accessory genes that contribute to the rapid adaptation and enhanced infectivity of the virus to a given host cell. Coronaviruses infect the human pulmonary system and cause widespread diffused alveolar damage, eventually causing hypoxia in the human body. Severe infection in immunocompromised individuals could lead to multiple organ failures and subsequent death.

Application of popular imaging-based techniques such as CT for the diagnosis of coronavirus disease during a pandemic situation is limited as they are time-consuming, costly, and nonportable in addition to lack of specific and reliable imaging markers. This chapter focuses on tackling these issues by taking two

The original version of this chapter was revised. A corrections to this chapter can be found at
https://doi.org/10.1007/978-981-15-6006-4_11

A. Ramadoss (✉)
Department of Research and Development, Nanolane, Le Mans, France

V. Raj
Faculty of Integrated Technologies, Universiti Brunei Darussalam, Bandar Seri Begawan, Brunei Darussalam

M. Kuniyil Ajith Singh
Research and Business Development Division, Cyberdyne Inc, Rotterdam, The Netherlands

P. Chandra, S. Roy (eds.), *Diagnostic Strategies for COVID-19 and other Coronaviruses*, Medical Virology: From Pathogenesis to Disease Control,
https://doi.org/10.1007/978-981-15-6006-4_5

distinct approaches. The first approach involves appropriate application of artificial intelligence to these imaging techniques leading to novel systems that provide reliable diagnosis that is time-efficient, thus rendering the methods suitable for rapid screening of the coronavirus disease. The second approach is to use an imaging technique such as ultrasound imaging that is immune to common problems faced by CT. Ultrasound (US) imaging is one of the most popular radiation-free medical imaging modalities with unprecedented spatial resolution and imaging depth. US technique offers high-resolution deep-tissue images comparable to that of CT scans. The lung US imaging can potentially identify changes in the physical state of superficial lung tissue (performance similar to CT imaging) making the modality suitable for use in the context of current pandemic. US imaging is simple, portable, user-friendly, time-efficient, and cost-efficient. This technique when combined with advanced methods like US Doppler, photoacoustic imaging, and AI-based algorithms holds strong potential in rapid screening of coronavirus disease such as COVID-19 even in remote and resource-limited settings.

Keywords

COVID-19 · Coronavirus disease · SARS-CoV2 · Imaging-based diagnosis · Portable devices · Rapid diagnosis · Artificial intelligence · Ultrasound imaging · Computer-aided diagnosis · Remote diagnosis · Tele-diagnosis · Radiology diagnosis

1   Introduction

Coronaviruses are known to infect a wide range of organisms across the species such as mammals, animals, and also amphibians (King 2012). In animals, they cause mild to severe upper and lower respiratory tract infections (Brian and Baric 2005). The first human coronavirus (H CoV229E) was reported in the 1960s that caused mild respiratory tract infections such as common cold in humans (Brian and Baric 2005). However, the epidemic of severe acute respiratory syndrome (SARS) (in 2002) drew worldwide attention when the responsible viral agent, SARS coronavirus (SARS-CoV), spreads across the globe and caused life-threatening acute respiratory syndrome (ARS), i.e., severe pneumonia with over 10% mortality (World Health Organisation 2003; Ksiazek et al. 2003). Since then, scientific understanding of coronaviruses has greatly increased, and currently over 100 species of coronaviruses have been discovered, and multiple druggable molecular targets have been identified, in addition to the development of numerous diagnostic tools to detect the pathogen (Luo et al. 2018; Mahapatra and Chandra 2020; Hu et al. 2017).

Nearly two decades later, in 2019, a coronavirus similar to SARS-CoV, called the severe acute respiratory syndrome coronavirus 2 (SARS-CoV2) with ~70% genomic sequence homology to the former is causing coronavirus disease (COVID-19), now a worldwide pandemic. As of May 2020, SARS-CoV2 has infected over three million people and has taken the lives of an estimated 285,000 thousand people on this planet (World Health Organisation 2020). The reason for such enhanced transmission and widespread is that SARS-CoV2 is highly contagious and airborne (Ksiazek et al. 2003; Bushmaker et al. 2020). This chapter attempts to provide a comprehensive overview of the pathobiology of coronaviruses including the pathological signature features observed in COVID-19 patients and further discusses the available imaging technologies for rapid detection of the disease in patients for use as portable devices and in remote settings.

2 Pathobiology of Coronavirus

Coronaviruses carry positive-sense single-stranded RNA (+ssRNA) as their genome, and they belong to the Coronavirinae subfamily in the Coronaviridae family. Human coronaviruses (HCoV) are further classified into groups B and C in the Betacoronavirus genus of the Coronavirinae subfamily (Fig. 1). The following sections will focus on HCoV. Readers are directed to other excellent reviews for detailed information on other subfamilies of Coronaviridae family (Balasuriya et al. 2017).

2.1 Structure of the Coronavirus

The major known structural proteins of coronavirus virions are, the spike (S), the transmembrane glycoproteins (M and E), dimers of membrane glycoprotein of hemagglutinin esterase (HE) found in few members of this family of viruses, and the Nucleocapsid protein (N). The virions of HCoV are spherical with a diameter between 80 and 220 nm (Brian and Baric 2005). These virions are enveloped with viral membrane glycoproteins (M and E) and are comprised of small "club-shaped" projections (~20 nm) called peplomers. The peplomers are composed of trimers of S glycoproteins (~200 KDa). Few exceptions of coronaviruses of this genus possess shorter spikes (~5 nm long) composed of dimers of hemagglutinin-esterase protein (~60 KDa per monomer) in addition to classic spikes such as the murine hepatitis virus (MHV), the bovine coronavirus, etc. (Brian and Baric 2005).

MHV has been studied extensively and is considered as a model organism that offers valuable insights into the molecular biology of coronaviruses (Brian and Baric 2005). The viral +ssRNA genome of the coronavirus is contained in an icosahedral structure composed of viral membrane glycoproteins M and E (~60 and ~9 KDa, respectively) as shown in Fig. 2a. The viral genome inside the icosahedral structure rests associated with nucleocapsid proteins (~50 KDa) and together they form a compact but loosely wound helical core (Hagemeijer et al. 2012).

2.2 Overview of the Coronavirus Genome

The genomic size of the +ssRNA HCoV ranges from ~27 to ~30 Kb and is largest among all of the known RNA viruses (Brian and Baric 2005, Almazán and González 2000; Lai and Cavanagh 1997). The nascent viral RNA genome is infectious as the +ssRNA acts as an mRNA template for the synthesis of RNA-dependent RNA

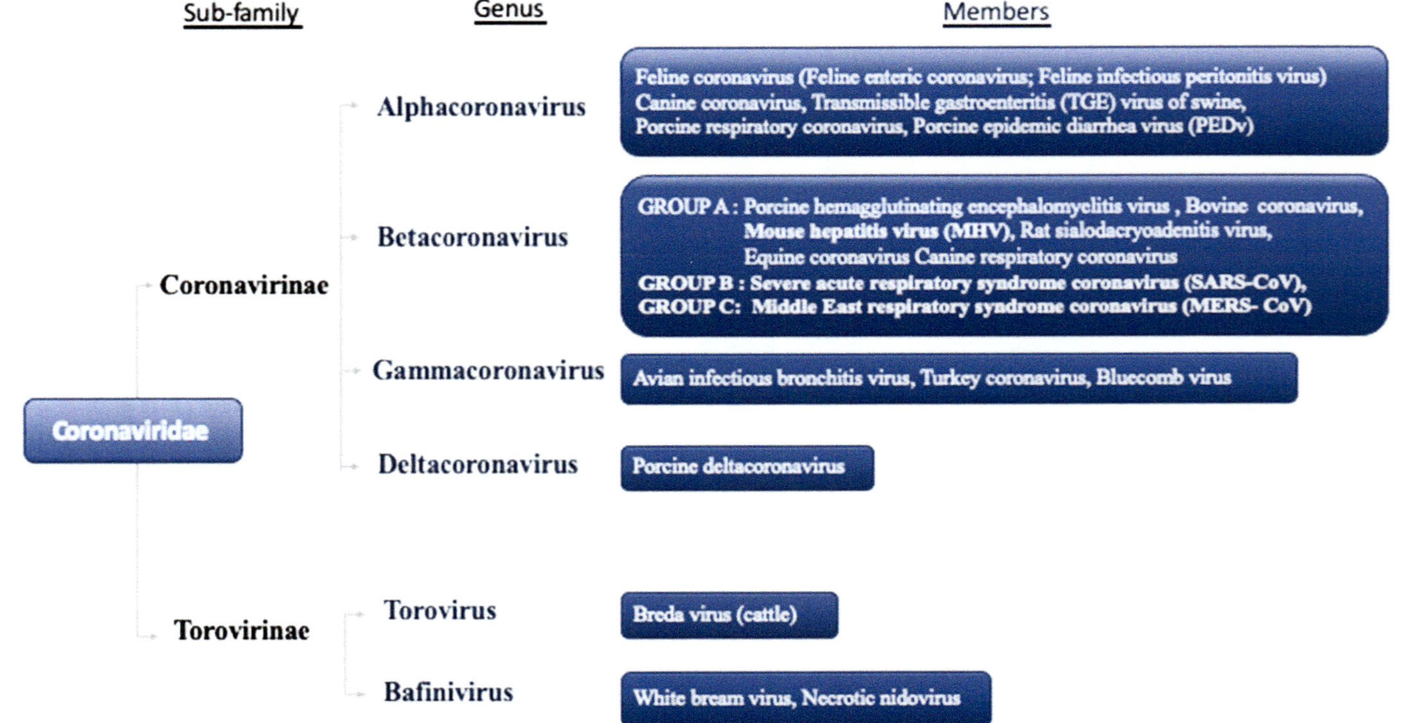

Fig. 1 Taxonomical classification of members of the Coronaviridae family: HCoVs are classified as group B (SARS coronaviruses) and group C (MERS coronavirus) in the Betacoronavirus genus of the Coronavirinae subfamily. The list of coronavirus members mentioned is not exhaustive. (Image adopted from Balasuriya et al. 2017 with permission)

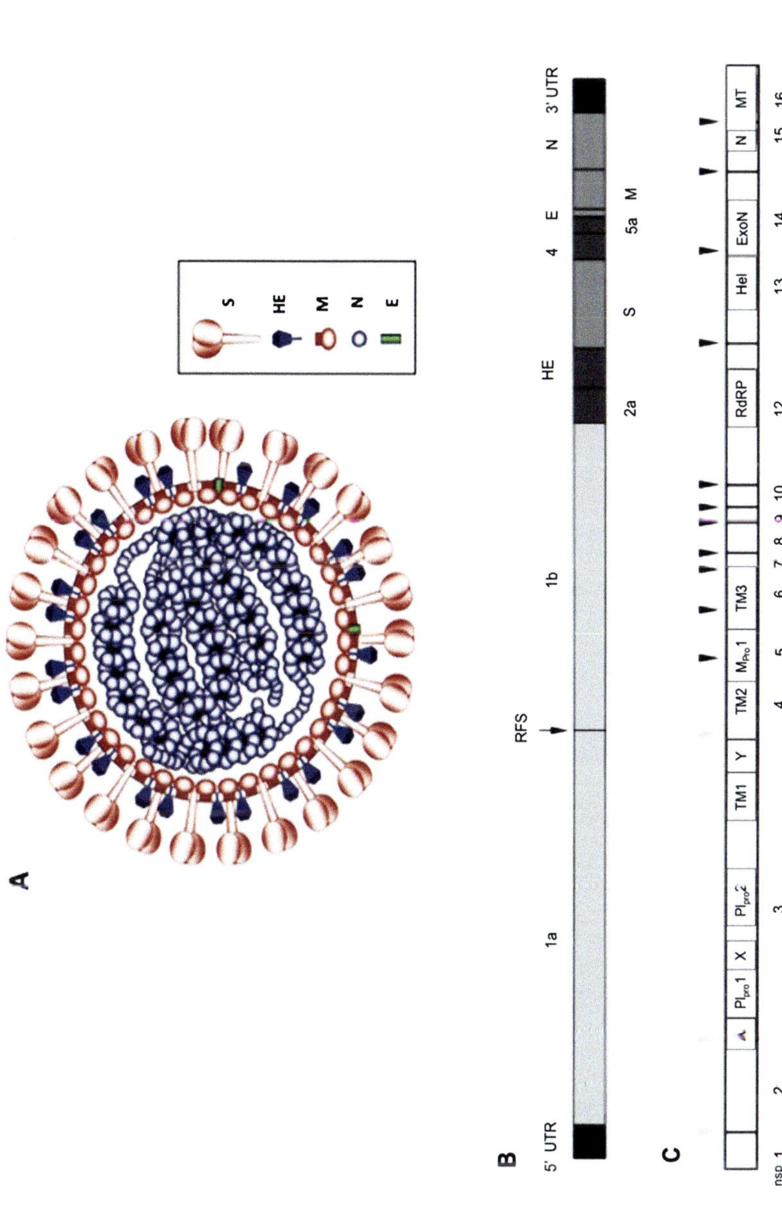

Fig. 2 Virion morphology and genomic structure of mouse hepatitis virus (MHV-A59): (a) structure of the virions of the members of betacoronavirus genus showing the arrangement of spike (S) glycoproteins, transmembrane proteins (M and E), and hemagglutinin esterase (HE). The RNA genome (i.e., nascent

polymerase (RdRP) upon entry into a host cell. The genome has a 5′ capped structure (i.e., methylated 5′ proximal end) and a poly-A-tail on the 3′ proximal end of their genome (Lai and Cavanagh 1997). A characteristic feature of the coronaviruses is that a large open-reading frame (ORF) (i.e., *gene 1*) encodes for a long polyprotein product that undergoes extensive proteolytic processing to generate over 16 non-structural proteins (nsps) (King 2012; Gordon et al. 2020; Deming et al. 2007). Figure 2b, c shows a schematic representation of the genome of MHV-A59 coronavirus and the polyprotein encoded by ORF 1ab displaying the proteinase cleavage sites on the polyprotein (pp1ab) to generate 16 mature nsps.

The *gene 1* localizes to 5′ proximal end (and is comprised of two ORFs: ORF 1a and ORF 1b) and encodes for viral proteins involved in RNA synthesis, genome replication, and proteolytic processing including the viral RdRP (Gordon et al. 2020). While the single gene that encodes for a huge polyprotein product of nsps spans nearly two-thirds of the viral genome, the genes that encode for key structural proteins such as spike glycoprotein (S), membrane glycoprotein (M), the envelope protein (E), and the nucleocapsid protein (N), an RNA-binding phosphoprotein localize to the rest of one-third of the genome toward the 3′ proximal end of the viral genome.

The viral genome contains numerous cis-acting regulatory elements and accessory genes that direct the expression of structural proteins, by spatially regulating the binding of RdRP to an ORF (Hagemeijer et al. 2012). The cis-acting regulatory elements allow the RdRP to transcribe certain ORFs in certain hosts based upon the availability of the ORF on the ssRNA in the host (Brian and Baric 2005). For instance, conserved slippery sequences (i.e., UUUAAAC), often found at the 5′ proximal end of the genome and on junctions of overlap of ORFs1a and ORFs 1b, form secondary stem-like loop or pseudoknots and, thus, regulate the availability of RdRP binding site for the particular ORF. They play an important role in the expression of the viral gene proteins involved in the genome replication and structural proteins (Brian and Baric 2005). Another set of conserved packaging

Fig. 2 (continued) +ssRNA) is stabilized by association with nucleocapsid protein (N) which forms the core of the structure. (**b**) Schematic representation of the genomic organization of MHV–A59 with 5′ methylated nucleotide cap and 3′ poly (A) tail including the untranslated regions (UTRs). Two-third of the genome consists of open-reading frames (ORFs) 1a and 1b which encode for two large polyproteins that contain the replicase and other nonstructural proteins required for viral genome synthesis, replication, and proteolytic processing, while the rest of one-third of the genome encodes for structural proteins in the order of S, E, M, and N, interspaced by several accessory genes as depicted in figure. (**c**) Schematic representation of the large polyprotein. This polyprotein undergoes proteolytic processing by viral proteinases that reside in nsp3. Black arrowheads indicate proteinase cleavage sites. This generates 16 mature proteins: hydrophobic domains (TM1, TM2, and TM3) in nsp3, nsp4, and nsp6 are considered as RNA (-modifying) enzymes: the RNA-dependent RNA polymerase (RdRP; nsp12), the helicase (Hel; nsp13), the exonuclease (ExoN; nsp14), the uridylate-specific endoribonuclease (N; nsp15), and the methyl transferase (MT; nsp16). (Image adopted from King 2012 and Hagemeijer et al. 2012 with permission)

sequences, found in certain members of coronaviridae family such as bafini corona-virus, MHV, etc. is required for the packaging of synthesized genetic material and other RNA proteins in the host cell (Brian and Baric 2005).

The presence of accessory genes is a characteristic feature of coronaviruses that contribute to the evolution of coronaviruses by allowing them to efficiently adapt to their new host environment (Brian and Baric 2005; Bandyopadhyay 2020; Wang et al. 2020). Another unique feature of the coronaviruses is the exonuclease activity of nsp13, unlike any other +ssRNA viruses. The nsp 13 (RdRP) is processed from the large polyprotein that is encoded by the ORFs 1a and 1b and has an exoribonuclease activity, a $3'$-$5'$ proof-reading activity. This proof-reading activity is essential for maintaining efficient viral replication fidelity given the large genome of coronaviruses (Becker et al. 2010; Denison et al. 2011). Other instances of cis-acting elements that regulate viral genome replication are the leader sequence present at the $5'$ proximal end of the viral genome and the transcription regulatory sequences primarily found upstream of each ORFs. Currently, there have been two different strains of SARS-CoV2 that isolated the L and the S strain, of which the L strain is considered to be more virulent causing severe infection in infected individuals (Bandyopadhyay 2020; Wang et al. 2020).

2.3 Viral Entry and Replication

The spike (S) glycoprotein of the coronavirus virion recognizes specific host cell receptors; for instance, the SARS-CoV recognizes ACE2 (angiotensin-converting enzyme 2), and MERS-CoV recognizes dipeptidyl-peptidase 4 (CD 26) on the host cells in order to facilitate viral entry into the host cell (Ge et al. 2013). The host cell proteases cleave off the S protein and activates viral fusion on the plasma membrane or an endocytosis-mediated viral entry into the host cell (Brian and Baric 2005; Hagemeijer et al. 2012; Knoops et al. 2008). Upon attachment of the virion to the host cell, the host cell undergoes rapid and dynamic cytoskeletal and membranous rearrangements (Ge et al. 2013; Ciulla 2020). This leads to the formation of numerous double-membrane vesicles (DMVs) and convoluted membranes (CMs) which are formed as a membranous network in association with the endoplasmic reticulum (ER) as shown in the electron micrographs in Fig. 3.

The viral genome replication takes place in the cell cytoplasm at the periphery of the ER–Golgi intermediate compartment (ERGIC) (Dalasuriya et al. 2017; Knoops et al. 2008; Ciulla 2020; Gosert et al. 2002). The +ssRNA genome of coronaviruses serves as an mRNA template and undergoes translation exploiting the host cell translation machinery, i.e., ribosomes, to produce the large polyprotein with RdRP, viral encoded proteinases, and the structural proteins including the accessory proteins. These proteins are required to form membrane-associated replicase-tran-scriptase complex (Gosert et al. 2002; Oostra et al. 2007). This replicase-transcriptase complex includes a viral RdRP (nsp 13) that synthesizes a -ssRNA strand of the viral genome and then copies it back into a +ssRNA; thus a full +ssRNA is generated using the replicase activity of RdRP in the

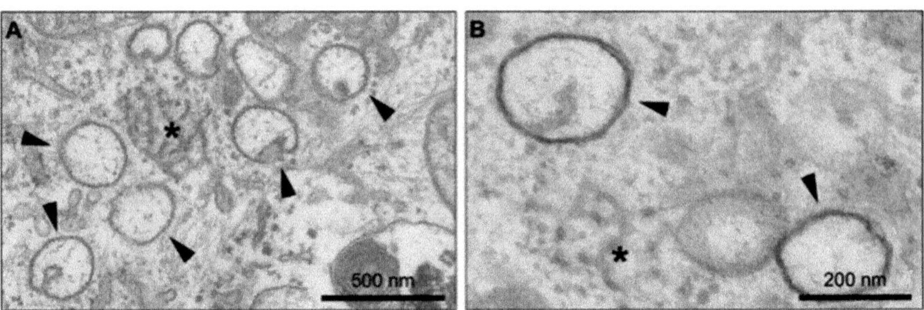

Fig. 3 Electron micrographs of coronavirus-induced membranous network: (**a**) Virion-induced DMV and CVs formed upon infection of SARS-CoV. (**b**) Higher magnification of DMV structures with double lipid bilayer. Arrows indicate DMVs and asterisks indicate CVs. (Image reprinted from Hagemeijer et al. 2012 with permission)

replicase-transcriptase complex (Balasuriya et al. 2017; Gosert et al. 2002). The coronavirus replication mechanism is detailed in the schema depicted in Fig. 4.

Following the genome synthesis, the viral structural proteins are synthesized by the RdRP in a discontinuous fashion. In addition to genome replication, the replicase-transcriptase complex also generates numerous discontinuous nested negative-sense RNA strands due to the presence of cis-acting internal regulatory elements (Balasuriya et al. 2017). These regulatory elements allow the RdRP to pause and translocate to the $5'$ end of the genome, where it extends the negative-sense RNA synthesized by copying the leader sequences (Balasuriya et al. 2017; van der Meer et al. 1999). The presence of the leader sequence allows the binding of the replicase-transcriptase complex and allows the nascent negative-sense RNA strand to be transcribed into segmented subgenomic RNA which further allows the expression of structural viral genes localized at the $3'$ proximal end of the viral genome. The post-translationally processed structural viral envelope proteins carry specific signals (such as O-linked or N-linked glycans) that direct these proteins toward ER and the budding Golgi bodies; hence these proteins localize at the interior of the DMVs and CMs. Subsequently, the packaging sequence (one of the cis-acting regulatory sequence) together with other nonstructural viral proteins and proteases and the viral particles are assembled into infectious virions inside the vesicles and are then released outside of the host cell by exocytosis (Balasuriya et al. 2017; Oostra et al. 2007; Jerome et al. 2004; Thackray et al. 2007). The coronavirus virion assembly mechanism is detailed in the schema depicted in Fig. 4.

2.4 Pathological Signature of Coronavirus Infection in Humans

HCoV viruses were known to cause the common cold and mild respiratory illness which seldom required medical attention. In adults, often symptoms of rhinitis, sore throat, and sometimes coughing were reported (Stephen et al. 2012). However, in infants, patients with asthma or chronic lung disease symptoms were reported worse

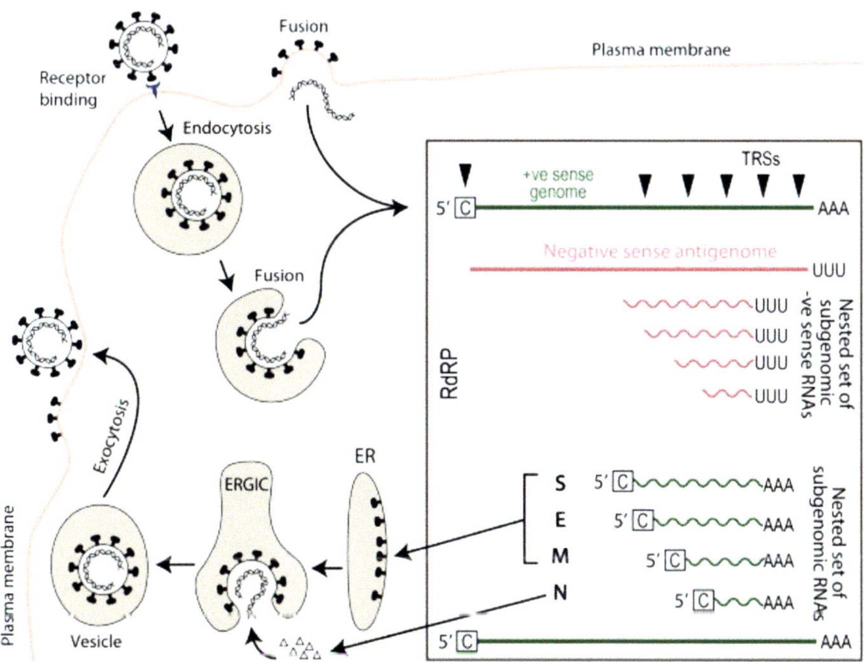

Fig. 4 Schematic overview of coronavirus entry, replication, and virion assembly in a host cell: Coronavirus replication primarily occurs when the viral S protein recognizes a specific host cell receptor and binds to it. The viral proteins along with the genome are endocytosed into the host cell. The endocytosis process initiates a massive of cytoskeletal and membranous rearrangements. The RNA-dependent RNA polymerase complex (RdRP) regulated by the transcriptional regulatory sequences (TRS) synthesizes viral structural proteins and performs genome replication using the replicase activity in the RdRP complex in the cell cytoplasm. Furthermore, via discontinuous synthesis, the RdRP also generates numerous copies of structural proteins. These viral proteins eventually are tagged that directs them to localize into DMVs and CMs. Eventually, the ER–Golgi intermediate compartment (ERGIC) internalizes these viral proteins which thus protects the material from host cell immune responses and allows the complete assembly of the proteins into active virions inside the vesicles which are then exocytosed out of the cell. (Image reprinted from Balasuriya et al. 2017 with permission)

as severe infection could cause bronchitis and pneumonia (Stephen et al. 2012). So far, the treatment for such infections has been symptomatic and supportive, and general hygiene including hand wash with standard alcoholic disinfectant has proven to be effective. However, the virus appears to have evolved over time with enhanced infectivity causing more serious infections in humans such as that observed during SARS and COVID-19 pandemic (Stephen et al. 2012).

The HCoV enters and infects the upper respiratory tracts where they target the respiratory tract parenchyma to replicate and, subsequently, spreads to the lower respiratory tracts. Eventually, the virus enters the gastrointestinal tracts and causes secondary infections (Stephen et al. 2012). SARS-CoV primarily infects the lower respiratory tracts from where it spreads to other parts of the body (Stephen et al.

2012). Recently, there is some emerging evidence that SARS-CoV2 is capable of infecting other body parts such as the brain, liver, stomach, and heart as well (Wang et al. 2020). The primary route of transmission of SARS-CoV is via contact with respiratory droplets and infected surfaces; similarly, that of SARS-CoV 2 is by inhaling the coughed out infectious respiratory droplets (particularly within 1 m distance of the infected person, as the pathogen appears to be airborne) and contact with the infected surface (Sahin et al. 2020). Most of the patients exposed to SARS-CoV2 develop symptoms with a median incubation time of 3 days, although the onset of symptoms could be anytime between 0 and 24 days depending on the exposure to the virus, the condition of the individual's immune system, and the severity of infection (Wang et al. 2020). The most common symptoms at the onset include fevers above 38 °C with chill and fatigue, runny nose, cough, and sore throat and with the possibility of gastrointestinal symptoms, i.e., diarrhea, and most often accompanied with characteristic shortness of breath that results in hypoxia (Wang et al. 2020). Often individuals with weakened immune systems or other chronic illnesses (10–20% of the patients) experience severe hypoxia and require the support of ventilators (Wang et al. 2020; Stephen et al. 2012).

Clinically, the pulmonary infection causes diffuse alveolar damage, which is considered as the first stage of damage. Specific host immune response results in cytokine dysregulation and results in large macrophage and T-cell infiltration on the respiratory tract parenchyma and causes type 2 pneumocyte proliferation. The infiltration of the immune cells can be observed on chest X-rays as patches (Wang et al. 2020). Subsequently, the virus infects the gastrointestinal tract and replicates in the host cell enterocytes. This results in diarrhea. The virus can thereafter spread to other body parts via contaminated body fluids such as urine, blood, etc. (Stephen et al. 2012).

Currently, suspected individuals undergo serological testing that checks for the presence of antibodies to a specific pathogen and molecular testing including polymerase chain reaction for pathogen detection. Covid-19 confirmed patients are further advised to undergo imaging-based diagnosis for the assessment of lung infection (American College of Radiology 2020).

3   Medical Imaging-Assisted Detection of Coronaviruses

Radiology is the primary tool exploited to assess the infected lungs and the severity of the infection and to discuss the course of medical treatment for patients with established coronavirus disease such as COVID-19. The following sections detail the importance and case studies of medical imaging and further discuss the new age cutting edge technologies such as the role of artificial intelligence (AI)-assisted computed tomography and introduce ultrasound imaging-based detection techniques.

3.1   Computed Tomography

Computed tomography or CT scan is a medical imaging procedure that uses multiple X-ray measurements taken from different angles at computer-processed combinations in order to produce cross-sectional images of specified regions of the object under scan (Goldman 2007). This provides optical image sections of the scanned region thus allowing us to visualize the object in a noninvasive manner (American College of Radiology 2020). The term "computed tomography" is most often synonymous with X-ray CT as it is the most commonly exploited form of CT. There are many other types of CT such as positron emission tomography (PET) and single-photon emission computed tomography (SPECT), computed tomography angiography (CTA), etc. CT allows the detection of lung pathologies with great precision and provides images at an unprecedented resolution. Thus, the technology offers highly valuable quality information to assess a lung condition. It can be used for detecting both acute and chronic changes in the tissue of the lungs.

Before the imaging procedure, a contrast dye (iodine-based contrast) is injected into the body for the better visualization of the object under investigation. During a CT scan, a loose-fitting outfit made of thin fabric (usually cotton) is provided to obtain an unperturbed scan with high resolution. Metal objects are avoided as they create artifacts on CT images. For CT pulmonary angiogram (CTPA), a noncontrast scan is taken prior to the scan with the contrast injection for better interpretation of the scan results (Bell 2020).

CT is extensively exploited in hospitals for its quality information that aids in clinical decision making. Albeit the aforementioned advantages, the non-mobility, the cost incurred including the technical resources for the equipment, and skilled personnel requirement render this technique unsuitable for the rapid detection and diagnosis of coronavirus disease (American College of Radiology 2020; Simpson et al. 2020).

3.1.1 Findings of COVID-19 on CT

Increasing CT reports of COVID-19 patients suggests that SARS-CoV2-induced pneumonia can be well distinguished from other viral pneumonias (Simpson et al. 2020; Chung et al. 2020; Kong and Agarwal 2020; Bernheim et al. 2020; Bai et al. 2020). COVID-19 patients typically present with CT findings that correspond to organizing pneumonia (i.e., histological patterns of alveolar inflammation with the presence of polyps in the alveolar ducts) (Simpson et al. 2020; Chung et al. 2020; Kong and Agarwal 2020; Bernheim et al. 2020; Bai et al. 2020; Pan et al. 2020). Correspondingly, the CT imaging feature of COVID-19 pneumonia includes peripheral and nodular GGO patterns (i.e., fuzzy opacity patterns that do not obscure the underlying structures) commonly observed as bilateral and multilobular, with or without the presence of consolidation (depending upon the amount of liquid filled in the region of compressible lung tissue is instead of air) in peripheral, posterior, and diffuse or lower lung zone distribution (Chung et al. 2020; Kong and Agarwal 2020; Bernheim et al. 2020; Bai et al. 2020; Pan et al. 2020; Salehi and Abedi 2020).

Although there are typical findings of COVID-19 in a large number of cases, a significant number of cases have opacities without a specific or clear distribution (Simpson et al. 2020; Bernheim et al. 2020). Lung abnormalities such as emphysema and diffuse parenchymal lung disease are typically observed in cases of viral pneumonia, which could also be associated with increased morbidity in case of

COVID-19 (Simpson et al. 2020). In addition, lymphadenopathy (abnormal size or consistency of the lymph nodes) and pleural effusion (build-up of excess fluid between the layers of the pleura outside the lungs) have also been reported in COVID-19 but are very rarely observed (Chung et al. 2020; Ng et al. 2020).

3.1.2   Categories of CT Imaging Features Related to COVID-19

Categorizing the CT imaging features related to COVID-19 pneumonia will allow rapid and accurate assessment of the infection, in addition to standardizing the reporting language for the disease. Based on the currently available COVID-19-related CT findings, Scott et al. classify them into four principle categories (Simpson et al. 2020). The following sections will elaborate these categories in detail.

3.1.2.1 Typical Appearance of COVID-19 Pneumonia

Based on current CT findings, there are at least three types of commonly reported GGO patterns that appear to be signature imaging features of the COVID-19 pneumonia. (1) Peripheral and/or nodular (i.e., mass-like) GGO patterns with or without consolidation that is often bilateral and multilobular, (2) multifocal and rounded GGO patterns with or without consolidation or visible intralobular lines, (3) reverse halo sign (i.e., central GGO surrounded by dense consolidation of more or less complete ring as observed on a high-resolution CT) or other signs of organizing pneumonia such as diffuse GGO patterns with linear, curvilinear, or perilobular opacities, and consolidation. As the abovementioned imaging patterns can be observed during other infection processes as well (i.e., they produce similar imaging patterns), it is worth mentioning here that the differential diagnosis includes influenza pneumonia and organizing pneumonia which can also be seen in drug toxicity and connective tissue disease (Simpson et al. 2020). Figures 5, 6, 7, and 8 are case studies of typical appearance of GGO patterns in COVID-19 patients in different age categories.

3.1.2.2 Indeterminate Appearance of COVID-19 Pneumonia

This category includes imaging features of COVID-19 pneumonia that are nonspecific to the diagnosis. A wide variety of diseases like hypersensitivity pneumonitis among many others present with overlapping features of varied GGO patterns, similar to that in Covid-19. For instance findings of multi-focal diffuse GGO patterns with unspecific distribution or very small non-rounded GGO patterns with non peripheral distribution are conjoining findings among different lung abnormalities and hence give inconclusive diagnosis of the disease. Thus, such CT findings are not sufficient for a confident COVID-19 diagnosis (Simpson et al. 2020) (Figs. 9 and 10).

3.1.2.3 Atypical Appearance of COVID-19 Pneumonia

CT findings that are typical of other diseases and are uncommon but reported in association with COVID-19 pneumonia are classified as atypical appearance. For example, the lobar or segmental consolidation observed in case of bacterial pneumonia, or the cavitation observed from necrotizing pneumonia (destruction of the underlying lung parenchyma resulting in multiple small, thin-walled cavities), or the tree-in-bud opacities (that indicates some degree of airway obstruction) with

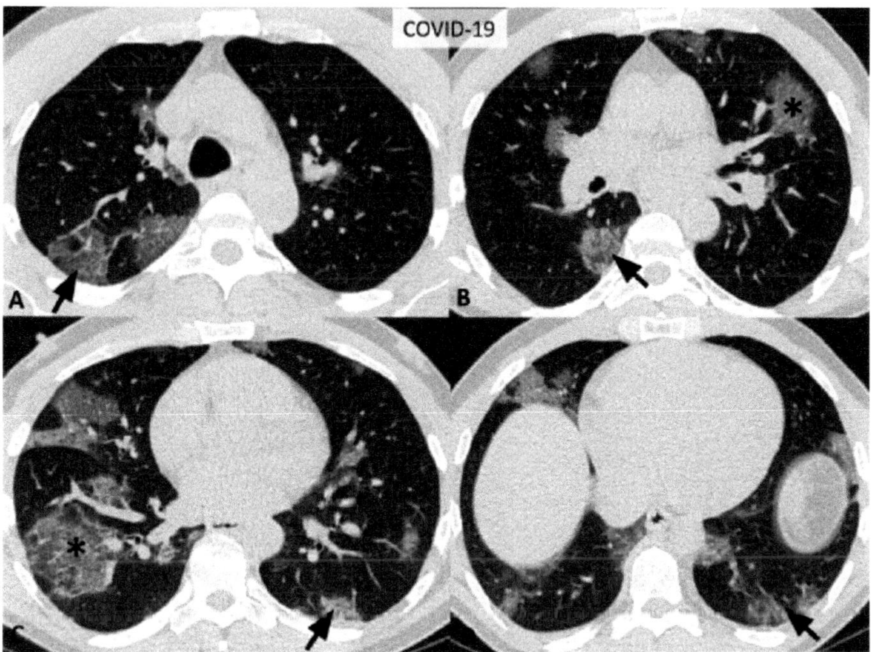

Fig. 5 Case study 1 of typical appearance of GGO patterns of COVID-19 pneumonia on chest CT: (**a–d**) Unenhanced, thin-section axial images of the lungs in a 52-year-old man with COVID-19 showing bilateral, multifocal rounded (asterisks), and peripheral GGO (arrows) with superimposed interlobular septal thickening and visible intralobular lines. (Image reprinted from Simpson et al. 2020 with permission)

centrilobular nodules (nodular opacity within the center of the secondary pulmonary lobule) observed in case of tuberculosis or mycobacterial infections, can be classified as atypical appearance for COVID-19 pneumonia. Figures 11, 12, and 13 show case studies as examples of COVID-19 patients with atypical CT imaging features. In these cases, although patients tested positive for COVID-19 disease, the lung damage observed on CT images is typical of another secondary infection (Simpson et al. 2020).

3.1.2.4 Negative for Pneumonia

CT imaging features of COVID-19 patients that are not suggestive of any pneumonia are classified as negative for COVID-19 pneumonia. The CT images show no signs of parenchymal abnormalities that are attributable to COVID-19 infection (Simpson et al. 2020). This is particularly observed when COVID-19 patients receive CT scans during early stages of the symptom onset. Such reports generate false-negative results and make it difficult for clinicians to determine the disease progression.

3.1.3 Pediatric CT

CT scan-based monitoring allows to assess and monitor the severity of the infection among the children as well. A study was conducted on five COVID-19-positive

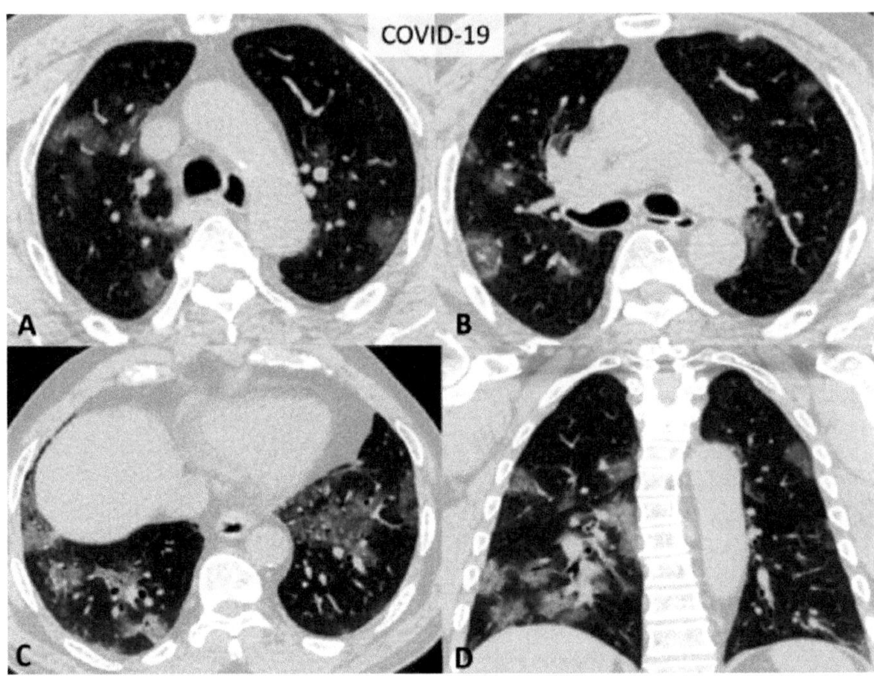

Fig. 6 Case study 2 of typical appearance of GGO patterns of COVID-19 pneumonia: Bilateral, multifocal rounded, and peripheral GGO patterns observed in a 77-year-old man with COVID-19. (**a–c**) Unenhanced, thin-section axial images, and (**d**) coronal multiplanar reformatted (MPR) images. (Image reprinted from Simpson et al. 2020 with permission)

children (tested positive based on RT-PCR results) who were subjected to chest CT scan. Only three among them displayed lung abnormalities, i.e., changes in their GGO patterns. As the children clinically recovered, these opacity patterns were found to have been resolved (Li et al. 2020a).

Similarly, another study was conducted on 171 children with COVID-19 in hospital settings. Abnormal GGO patterns were observed in one-third of the children, while over 16% of the population did not present with any pattern indicating pneumonia (i.e., the images were negative for COVID-19 pneumonia) (Rubin et al. 2020). However, for the rest of children, the CT findings observed appeared similar to adults as most of the GGO patterns appeared as bilateral patchy opacities, but less florid.

3.2   Chest X-Rays

Chest X-ray is one of the most widely used imaging techniques and is extensively used to diagnose lung diseases. Often physicians request chest X-ray in order to assess the status of pulmonary infection after the confirmation of COVID-19 through the molecular-based diagnosis. The findings of chest X-ray in the acute phase usually detect sticky mucus formations in one or both lungs. Chronic stages of the

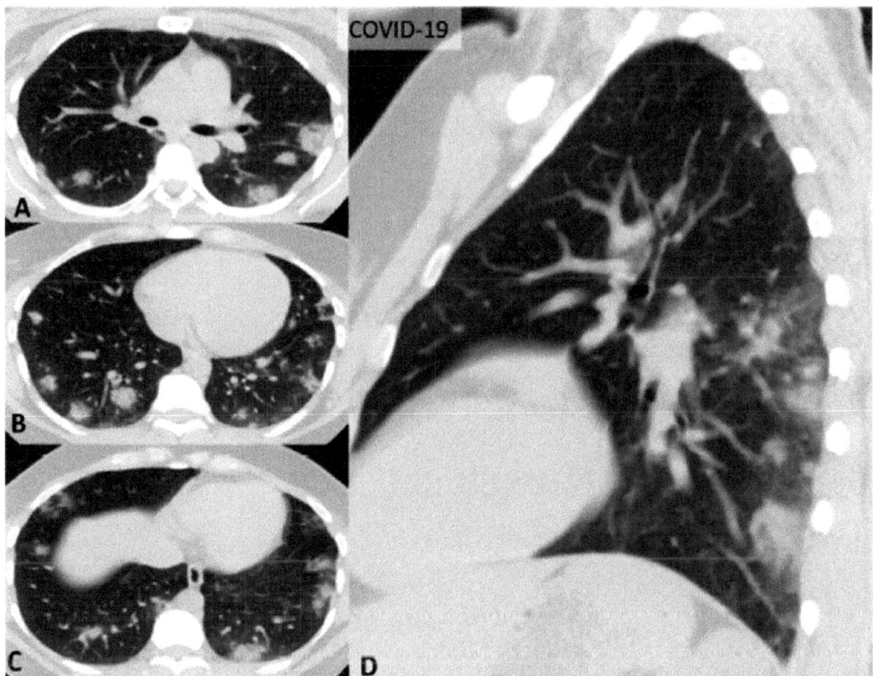

Fig. 7 Case study 3 of typical appearance of GGO patterns of COVID-19 pneumonia on chest CT: (**a–c**) Unenhanced axial images and (**d**). Sagittal coronal multiplanar reformatted (MPR) images of the lungs in a 29-year-old man with COVID-19 disease showing multiple bilateral, rounded consolidations with surrounding GGO. (Image reprinted from Simpson et al. 2020 with permission)

disease include "multifocal consolidations" (i.e., consolidation refers to the filling of pulmonary spaces with fluid or exudates) and "pleural effusions" (i.e., a collection of fluid in the pleura which is the outer layer of the lungs) (Kooraki and Hosseiny 2020; Staff 2020). Viral testing based on molecular methods is the current golden standard for the detection of COVID-19. However, Chest X-ray is recommended for COVID-19 patients with severe difficulty in breathing (Staff 2020).

The advantages of using a chest X-ray include its portability, speed, cost-effectiveness, and easy to disinfect the equipment. Portability and speed of the equipment and technology help to reduce waiting time and extra movements for the patients which is an important parameter for consideration during the current pandemic. Chest X-rays can be performed through the glass window of the patient's room which significantly decreases the exposure of medical staff to the virus (Rubin et al. 2020). This is particularly useful in resource-limited settings. Although the portability of the technique, rapid testing, and facile cleaning of the equipment are key advantages of this technique, significant false-negative results can be treated as a potential disadvantage for this technique.

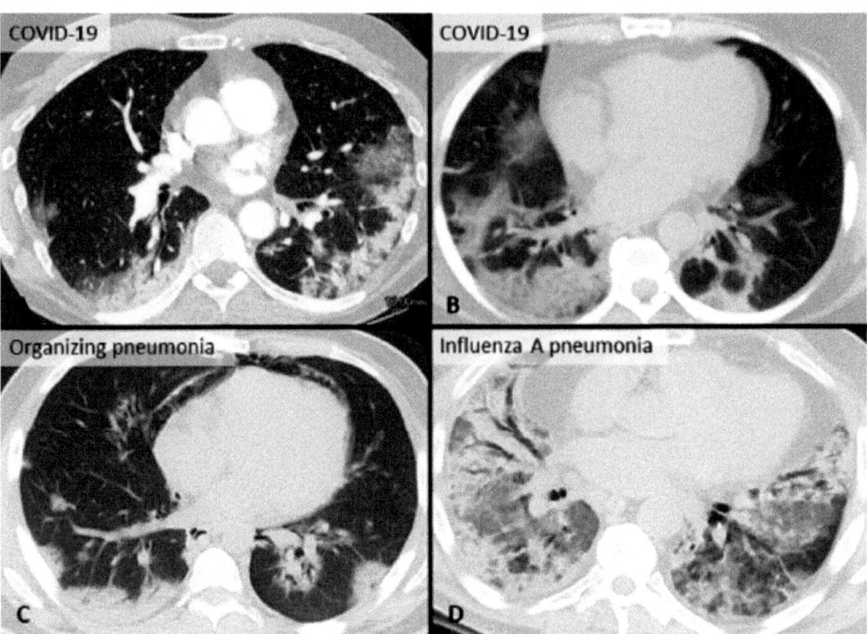

Fig. 8 Case study 4 of typical appearance of GGO patterns of COVID-19 pneumonia and other diseases with similar findings on chest CT: (**a–b**) Posterior, peripheral, and rounded GGO and consolidation in axial images of COVID–19 patients, (**c**) imaging features that correspond to organizing pneumonia secondary to dermatomyositis, (**d**) imaging features of influenza A pneumonia. Organizing pneumonia and influenza pneumonia can be indistinguishable from COVID-19 by CT. (Image reprinted from Simpson et al. 2020 with permission)

3.3　MRI Scanning

Magnetic resonance imaging (MRI) is a medical imaging technique that uses radio waves in the presence of a strong magnetic field to produce detailed images of tissues and organs. This technique facilitates the identification of abnormal tissues and changes in blood vessel structures and/or spinal cord. It is highly recommended for those patients who need imaging of joints, heart vessels, and brain pathology. Common indications include stroke, aneurysms, spinal code injuries, ligament injuries, multiple sclerosis, etc. In the current context of the COVID-19 pandemic, this technique is reserved for use only in emergency situations as recommended by the American College of Radiology (ACR) (American College of Radiology 2020). The bulkiness of the equipment and the inability to properly disinfect the machine raise concerns for contamination through contact (Murray et al. 2020). It is recommended to wear class 2 or 3 filtering facepiece respirator or an N 95 respirator for healthcare workers (which includes MRI staff members) as well as the patient to prevent the possible contamination. The presence of ferromagnetic components in the respiratory masks makes it unsafe for use by patients as well as healthcare workers and thus compromises safety against the virus during MRI scanning (Murray et al. 2020).

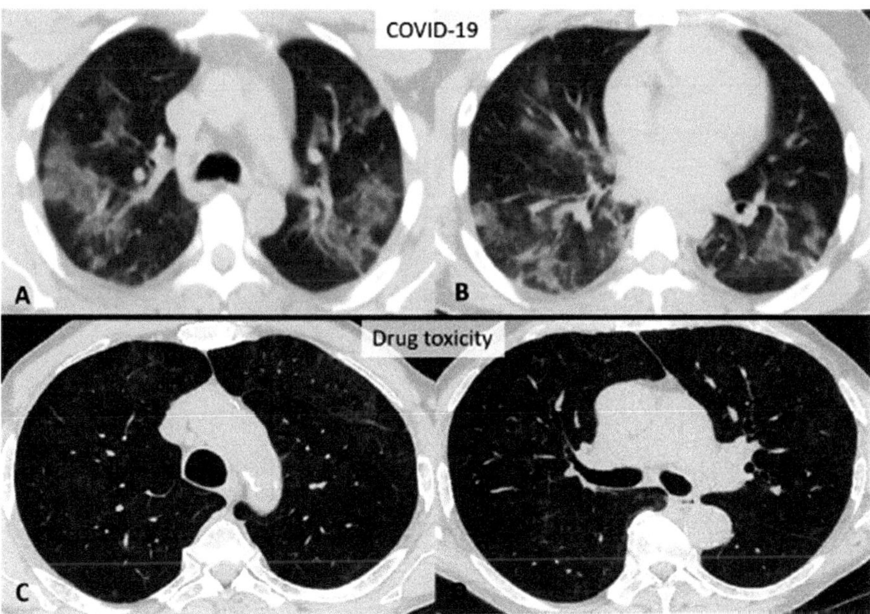

Fig. 9 Case study 1 of indeterminate appearance of GGO patterns of COVID-19 pneumonia: (**a, b**) Unenhanced axial images of two patients displaying patchy GGO patterns with non-rounded morphology without specific distribution. (**c, d**) Similar GGO patterns observed in cases of acute lung injury from possible drug toxicity. (Image reprinted from Simpson et al. 2020 with permission)

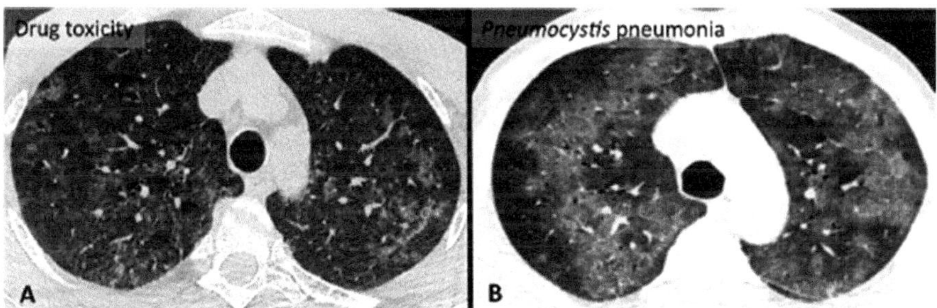

Fig. 10 Case study 2 of indeterminate appearance of GGO patterns of COVID-19 pneumonia: Unenhanced axial images of two individual patients with **a**. lung injury upon possible drug toxicity and **b**. pneumocystis pneumonia. The images show widespread GGO patterns with non-rounded morphology, and with no specific distribution. Similar patterns are also documented in some COVID-19 cases. (Image reprinted Simpson et al. 2020 with permission)

3.4 Limitations of the Popular Imaging-Based Techniques for Use in the Context of Coronavirus Disease Pandemic

Adherence to infection control protocols designed to minimize the risk of virus transmission and protection of healthcare professionals can be challenging while

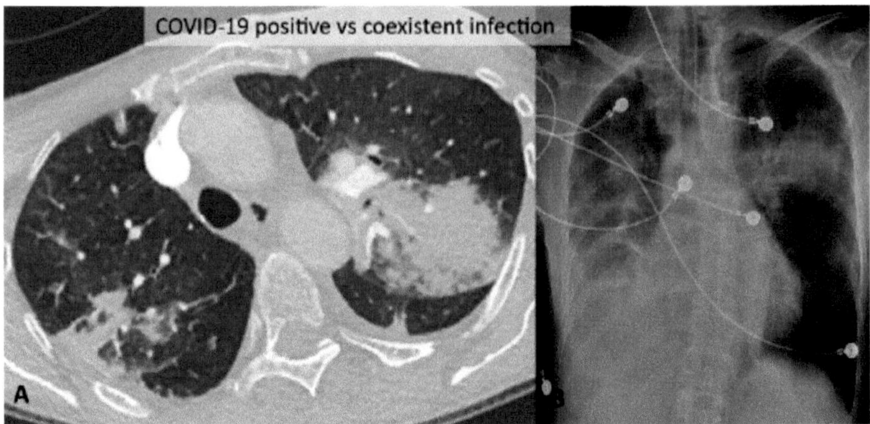

Fig. 11 Case study 1 of atypical appearance of COVID-19 pneumonia on chest CT: (**a**) Contrast-enhanced axial CT images, (**b**) frontal chest radiograph showing segmental consolidation without significant GGO. The damage of the lungs observed here could also be due to the presence of a secondary infection. (Image reprinted from Simpson et al. 2020 with permission)

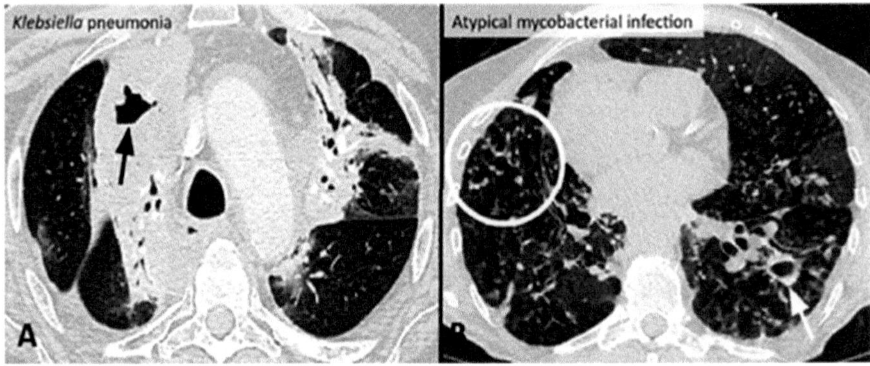

Fig. 12 Case study 2 of atypical appearance of COVID-19 pneumonia on axial chest CT images of two patients **a**. with *Klebsiella pneumonia* showing cavitation (arrow) and **b**. tree-in-bud opacities (circle) and a cavity (arrow) in nontuberculous mycobacterial infection. (Image reprinted from (Simpson et al. 2020) with permission)

using CT (Mossa-Basha et al. 2020; Kooraki et al. 2020; Rubin et al. 2020). As the imaging needs to be performed within an infected patient's isolation room, equipment portability is preferred, which is another factor that may favor CXR over CT in selected populations (Rubin et al. 2020). Apart from the issues mentioned above, the lack of unique imaging markers that allow confident classification coronavirus disease renders these imaging-based techniques unsuitable for rapid screening and diagnosis of the coronavirus disease such as COVID-19 (Staff 2020).

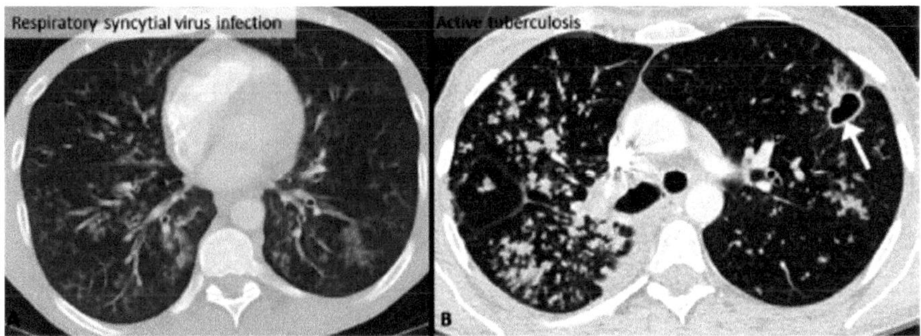

Fig. 13 Case study 3 of atypical appearance of COVID-19 pneumonia on axial chest CT images of two patients. Tree-in-bud opacities and centrilobular nodules are observed in **a**. lung infection caused by respiratory syncytial virus and **b**. lung infection caused by active tuberculosis with a small cavity visualized. (Image reprinted from Simpson et al. 2020 with permission)

3.5 Computer-Aided Approach

Application of popular imaging-based techniques such as CT for the diagnosis of coronavirus disease during a pandemic situation is limited as they are time-consuming, costly, and nonportable, in addition to lack of unique imaging features that allow confident diagnosis of the disease. This chapter focuses on tackling these issues by discussing some of the computer-aided approaches. The following subsections discuss teleradiology, appropriate application of artificial intelligence to these imaging techniques leading to novel systems that are time-efficient rendering the methods suitable for rapid diagnosis of coronavirus disease. Finally, this chapter introduces the readers to ultrasound imaging that is immune to common problems faced by CT and other popular imaging-based approaches.

3.5.1 Teleradiology for Remote Diagnosis

Teleradiology refers to the practice of a radiologist interpreting medical images while not being physically present in the location of diagnosis. Through teleradiology, the radiologist reads images and communicates with referring physician, without exposure to the work station. Teleradiology includes mobile applications that give the opportunity for radiologists and physician to interpret images and diagnoses and share reports via the cloud-based PACS (Picture Archiving and Communication System) from different locations (Chopra 2020) Teleradiology plays an important role, as it allows healthcare professionals to work remotely without being exposed to the disease-causing agent.

3.5.2 Artificial Intelligence-Assisted Novel Imaging Approaches

The approach to detect COVID-19 is similar to SARS on the frontlines; however, there is one big difference: the emergence of a powerful new weapon called artificial intelligence (AI) over the past couple of decades (Farid et al. 2020; McCall 2020).

Data mining plays a key role in biomedical sciences which allows predictions to identify and characterize pandemic with high accuracy (Farid et al. 2020). Artificial intelligence (AI) can be defined as an ability of machines to understand and learn from new experiences and respond to new inputs, in addition to carrying out specific tasks in an autonomous manner. In the context of the COVID-19 pandemic, AI can have a significant impact in different areas, for instance, from monitoring the disease progression, detection, and diagnosis including detecting imaging markers for disease prognosis to drug discovery research.

Key factors that dictated appropriate preventive measures during the international SARS response in 2003 are the definition and characterization of the causative organism, historical history of infection including incubation and mortality levels, knowledge of transmissibility, and the target populations. The WHO suggests a similar approach for the current pandemic, but this time additionally, the WHO calls for the exploitation of intelligent computer programs which could be specifically designed and developed to check the current outbreak of SARS-CoV-2. These semi-autonomous intelligent programs should be supplied with previously available knowledge available of the SARS epidemic, in addition to the abovementioned key factors about the novel coronavirus, thus ensuring that AI is linked with evidence information from a variety of sources including social media (Farid et al. 2020; McCall 2020).

The following are some instances where AI has been exploited for analyzing CT images:

(1) Prediction of the effect of seasonal influence on COVID-19 occurence: The AI-based algorithm predicts how the warmer weather will benefit the spread of the virus. (Farid et al. 2020).
(2) Rapid detection of COVID-19 lung infection from pulmonary radio-images: The AI-based algorithm distinguishes pulmonary images of COVID-19 lung infection from other respiratory infections within ~10 s compared to traditional manual diagnosis. The program identifies lesions of potential coronavirus pneumonia to quantify its length, shape, and distance and to analyze improvements in several lung lesions from the image, both of which aid physicians in making rapid decisions (McCall 2020). As the number of tests increases, the algorithm knows and increases its sensitivity along the process. Thus, AI can efficiently aid in rapid clinical diagnosis as well as clinical decision making.

3.5.2.1 Deep Learning-Based Algorithms

Deep learning (DL) algorithms permit the extraction of detailed information from CT images for coronavirus disease diagnosis as it provides deeper insight into the diagnosis. DL algorithms can derive graphical features from the patient's images, i.e., they can determine inconsistencies in a patient with positive corona lab report with no clinical symptoms and, thus, provide prepathogenic clinical examination (Farid et al. 2020; Zhu et al. 2020a; Li et al. 2020b; Cohen and Normile 2020). In the following paragraphs, some studies explaining the AI approaches have been detailed.

The following study proposes an AI algorithm with over 90% accuracy in distinguishing CT images of pneumonia of COVID-19 patients from other types of pneumonia. Over 250 CT images of COVID-19 confirmed cases, including those clearly identified as normal viral pneumonia and SARS, were studied (images obtained from the Kaggle database), and general procedure adopted is described in Fig. 14. Figure 15 describes in detail each step involved in the entire workflow. Four hybrid filter extraction methods that employ four different filters such as MPEG-7 edge histogram filter with Gabor filter, the pyramid of rotation, invariant local binary pattern histograms, and fuzzy 64-bin histogram that analyzes low-level image features, extracts features, and offers statistical hypothesis, respectively, were used to achieve high accuracy in prediction and improve the feature extraction methods (Li et al. 2020b; Farid et al. 2020a, b, c). Subsequently, hybrid feature extraction, i.e., selection of specific features for model building (variables, predictors) in machine learning, was done using composite hybrid attributes selection method (CHFS) with hybrid classifications for mixing multiclassifiers in order to enhance the study (Escudero et al. 2012). The helpful attributes that allowed to distinguish between the images were tested beforehand, and the approach was employed in conjunction with classifiers used by the stack hybrid classifications and convolution neural networks (CNN). The results obtained were finally compared with the CNN-based results. The program predicts the output after the application of 10-fold cross-validation with an accuracy of 96.07% compared to 94.11% accuracy using CNN-based classification. Figure 16 shows the prediction steps of the proposed system with a hybrid classification mechanism (Farid et al. 2020). To summarize, the proposed model performs better than the traditional classification methods because of the improved feature selection and enhancement of the classification mechanism and in addition to efficiently reduced wrong-negation rate (Farid et al. 2020). Such algorithms demonstrate that AI can be efficiently used for the timely and accurate diagnosis of COVID-19, with radiological characteristics.

In another study led by Fei Shan, the authors developed an automated classification and quantification method based on DL to analyze infectious regions as well as whole chest CT scans for the progression of lung disease (Shan et al. 2020). This method allows continuous monitoring of infectious lung areas, in addition to reliably measuring their shape, volume, and percentage of infection (POI) in CT scans of COVID-19 patients. The method uses a technique called human in the loop (HITL) to iteratively produce training samples in order to define hundreds of COVID-19 CT

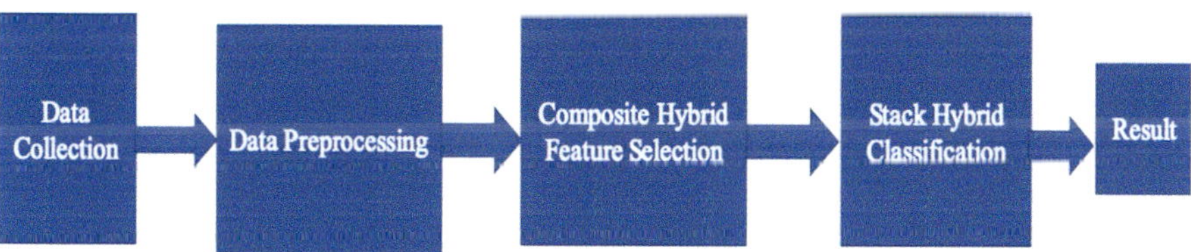

Fig. 14 The proposed layout for the clinical dataset analysis. (Image adopted from Farid et al. 2020 with permission)

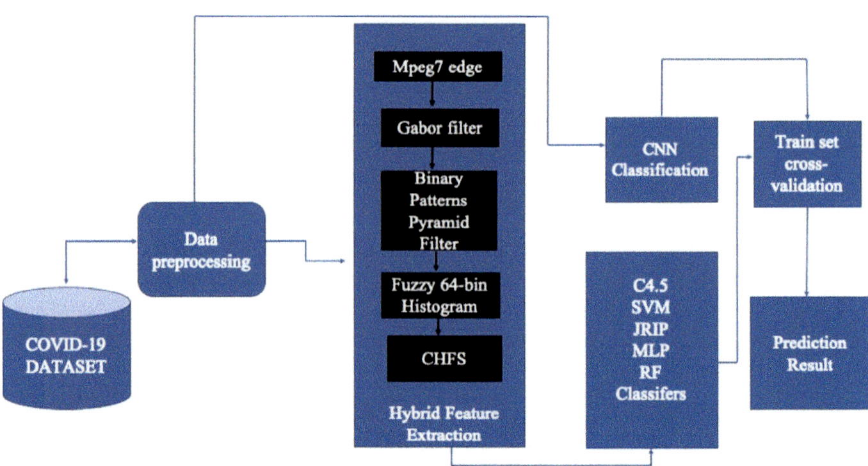

Fig. 15 Proposed extraction and classification of features for COVID-19. (Image adopted from Farid et al. 2020 with permission)

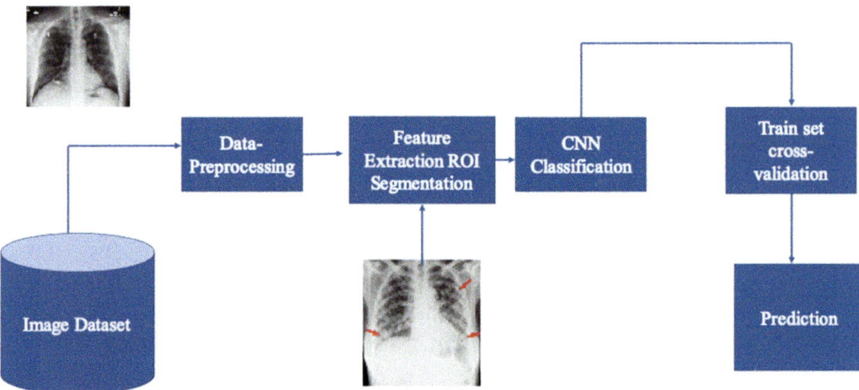

Fig. 16 Shows the prediction model for the proper classification of CNN CT images. (Image adopted from Farid et al. 2020 with permission)

images, as depicted in Fig. 17. This is the first study to employ the HITL strategy to detect COVID-19 infections on CT scans. The authors built a DL segmentation method in order to quantitatively evaluate the pathogens. This approach requires radiologists to efficiently interpret the effects of DL segmentation and iteratively introduce more samples to refine the model so that the algorithm's output is signifi-cantly improved. The DL-based segmentation uses the "VB-Net" neural network to segment infectious regions on CT scans. The detailed workflow of this DL-based algorithm is shown in Fig. 18. Over 249 CT images of COVID-19 patients were used for training the program, while another 300 CT images of COVID-19 patients were

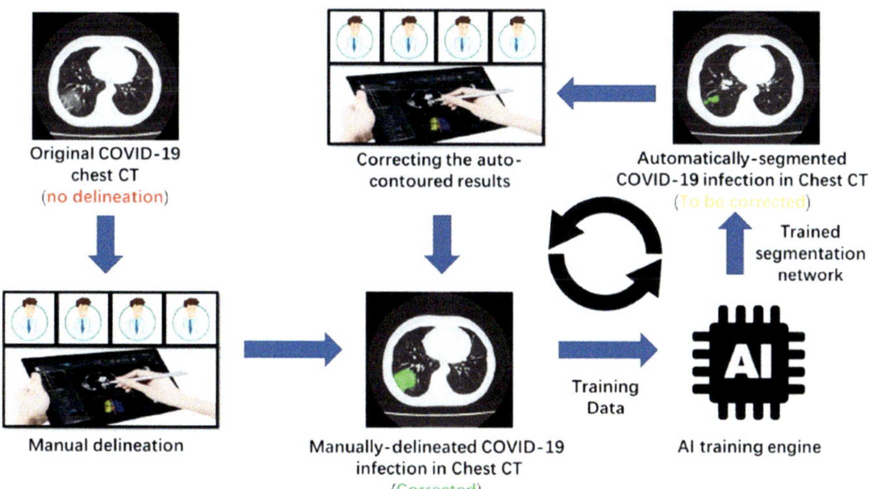

Fig. 17 The human-in-the-loop workflow. (Image reprinted from Shan et al. 2020 with permission)

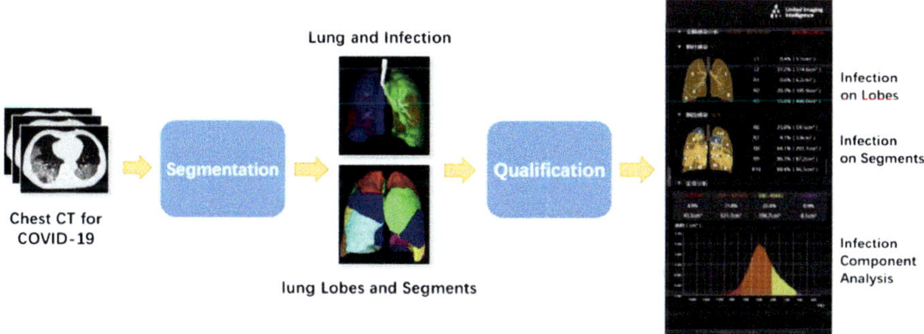

Fig. 18 Detailed workflow to identify the COVID-19 disease: Chest CT scan is fed into the DL-based segmentation system which uses the "VB-Net" neural network to segment infectious regions on CT scans. Quantitative metrics are then calculated to characterize infection regions on the CT scan, including but not limited to infection volumes and POIs in the entire lung, lung lobes, and bronchopulmonary segments. (Image reprinted from Shan et al. 2020 with permission)

tested. HITL technique allows radiologists to optimize the automated classification of each event in order to improve the manual delimitation of training CT pictures. In order to determine the efficiency of the DL-based method, the dice similarity coefficient, the volume, and POI discrepancies are determined between the automated and manual segmentation results of the validation collection. The abovementioned method provided 91.6% $\perp$ 10.0% dice similarity coefficients between automated and manual sections and an average 0.3% POI error estimate for the whole lung on the testing data set. Compared to cases that are manually categorized which could take up to 5 h, the HITL approach performs this categorization within minutes (takes up to 4 min). The quantitative evaluation shows high precision of POI metrics for the automatic infection area (Shan et al. 2020).

AI-based approaches are often employed where black boxes that have the capacity to learn from their environment, i.e., learn from inputs from clients are used. A new feature of the abovementioned approach is the use of the HITL technique in the training of the segmentation network (as described in the previous study). A large number of CT data annotations require AI-based approaches for automated quantitative evaluations, as the compilation of data annotated is quite expensive or even difficult. DL-based HITL technique makes the annotation process quicker. HITL technique allows radiologists who are acquainted with the AI method to engage in the training phase. This dynamically integrates radiologist's professional knowledge into the approach (Shan et al. 2020). Currently, multiple research focuses on such an automated DL-based segmentation approach to quantify and correlate imaging parameters with syndromes, epidemics, and treatment therapies, which may eventually offer deeper insights into imaging markers and provide results for better diagnosis and care for coronavirus-induced diseases such as COVID-19 (Shan et al. 2020).

In the wake of the COVID-19 pandemic, multiple AI-based approaches have been employed for the rapid detection of viral pneumonia induced by SARS-CoV2 and have been successfully commercialized (CareMentor 2020). One such commercialized approach uses a special advanced computer vision developed in-house that is based on artificial neural networks to detect pneumonia on images of chest radiographs and CT scans (CareMentor 2020). These systems are used to identify lung disorders and malignancies on chest X-rays and on CT images. The model has been shown to outperform the detections made by qualified radiologists (CareMentor 2020). Such AI-based approaches will drastically improve the detection and diagnosis of lung conditions in short notice as new coronavirus disease emerges among a population.

3.6 Imaging-Based Point-of-Care Devices

Point-of-care (PoC) device, based on the definition of the WHO, is a device that should meet the ASSURED criteria: affordable, sensitive, specific, user-friendly, rapid and robust, equipment-free, and deliverable. In other words, PoC tests would have to offer a very cost-effective tests and efficiently provide trustable quality results that will aid in rapid clinical decision-making, thus dramatically increases the scope of diagnosis in communities with limited detection facilities and rapidly growing demand for the tests. The need for rapid PoC testing devices during a pandemic increases manifold times. This has been well echoed by the WHO, as the world faces a pandemic of SARS-CoV2-induced COVID-19 disease that has currently infected more than three million people around the globe and caused over 285,000 deaths worldwide (World Health Organisation 2003; World Health Organisation 2020).

Technological advancement particularly in the domain of nanophotonics and optical imaging has significantly improved the quality of healthcare in the last couple of decades (Boppart and Richards-Kortum 2014). Numerous detection techniques that are based on sensing biophysical, biochemical, and/or biological properties exist; however optical imaging-based technologies offer image-based data, in addition to multiplexed real-time imaging with a high temporal and spatial resolution (provides imaging of cells, tissues, and even molecules), and offer cost-effective rapid screening (miniaturized and portable) for operation in remote settings (compatibility to mobile phones and network platforms) and, thus, provide valuable inputs for rapid diagnosis and aid in clinical decision making. In the following sections, we have discussed the potential of ultrasound-based detection as a portable imaging device that is relevant in the context of the current pandemic situation.

3.6.1   Introduction to Ultrasound Imaging

Ultrasound (US) imaging is one of the most popular medical imaging modalities with unprecedented spatial resolution and imaging depth. In this PoC real-time imaging technique, US waves are transmitted into the tissue, and the reflected echoes are used to generate tissue acoustic attenuation maps that provide valuable structural details of the tissue (Sazbo 2004) (Fig. 19). US imaging can also assess the blood flow information with high resolution by a technique called as US Doppler imaging, which measures changes in the frequency of the US echoes and, thus, evaluates the movement of blood cells. Rich structural information from conventional pulse-echo images and quantitative functional temporal information (i.e., blood flow information) from the same US Doppler machine has demonstrated to be helpful for diagnosis and treatment monitoring of a wide range of diseases (Sazbo 2004; Soldati et al. 2020). US imaging technology is mature, ultraportable, and inexpensive; and thus, it is successfully commercialized by many companies as affordable machines for multiple musculoskeletal, rheumatology, cardiovascular, and anesthesia/intensive care imaging applications.

3.6.2   Ultrasound-Assisted Diagnosis of Lung Pathologies

Lung US imaging offers excellent diagnostic accuracy in diagnosing pleural effusion, pneumothorax, pulmonary venous congestion, and consolidation (Soldati et al. 2020). Given that lung US imaging can potentially identify variations in the physical status of subsurface lung tissue (performance similar to CT imaging), the role of lung US imaging in the context of the coronavirus disease such as the COVID-19 pandemic is highly relevant. It is worth mentioning that the capability of US imaging in detecting lung lesions well before the development of hypoxemia has already been demonstrated in experimental studies. Based on recent clinical findings from around the globe and experiences from other diseases (ARDS, flu virus pneumonia, etc.), lung US imaging may offer the following key advantages over other gold standard imaging modalities used for coronavirus disease screening (Soldati et al. 2020).

- Rich structural and functional details of lung health in a point-of-care remote setting

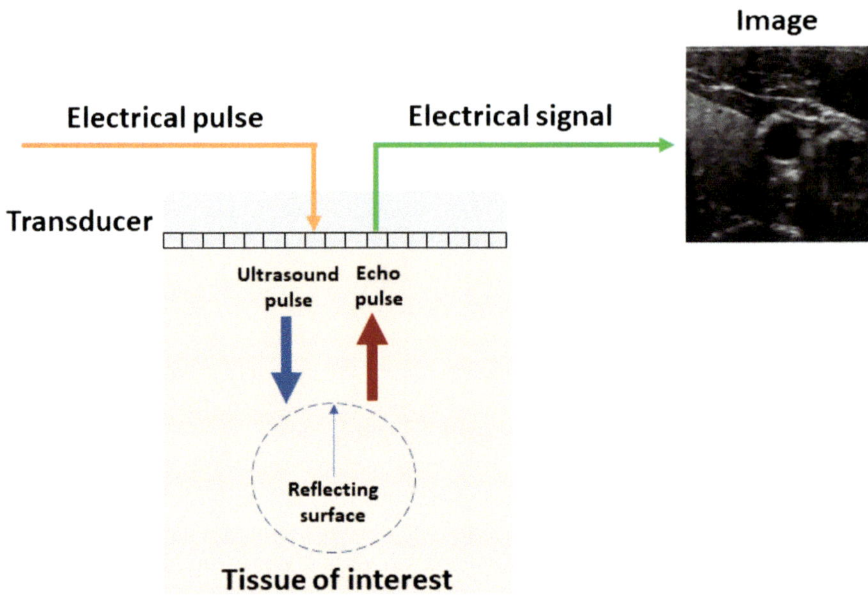

Fig. 19 Basics of ultrasound imaging in which ultrasonic pulses are transmitted to the tissue of interest, and the reflected echoes are used to generate an image representing an acoustic attenuation map

- Possibility to image symptomatic patients (with or without pneumonia) at their residence before admitting to the hospital
- Portability of the technology allows its easy exploitation in an emergency department or an ICU setting (US imaging systems can be easily used when the patient is connected to a ventilator)
- Possibility to decrease the number of healthcare professionals exposed to the virus during initial patient screening (imaging can be performed by a single clinician)

3.6.2.1 Detecting Pneumothorax

One of the key advantages of US imaging in the context of patients with coronavirus-induced SARS is the capability to define the changes affecting the ratio between tissue and air in the superficial lung (Faqin et al. 2020). In healthy individuals, the lung surface primarily consists of air, and the visceral pleural plane acts as a perfect acoustic reflector resulting in total reflection of incident US waves. High-intensity echoes from the pleural layer traverse back and forth between the US probe surface and chest wall resulting in line-like artifacts with brightness reduced over depth. However, in affected individuals, the scattering of US waves produces artifactual images characterized by horizontal reverberations of the pleural line (A-lines) and mirror effects. These reverberation artifacts are easy to differentiate as the distance between skin surface to pleural layer will be equal to the distance between different A-lines underneath it. The A-line is created by an intact "dry" lung parenchyma containing the air combined with normal lung sliding. When the sliding lung is absent, it is a strong indication of pneumothorax, which is a common condition currently observed in COVID-19 patients (Buonsenso et al. 2020).

3.6.2.2 Detecting Lung Interstitial Syndrome

If there is acoustic impedance mismatch because of reduction in the ratio between air, tissue, fluid, and/or other biological components, the lung may not reflect the US completely. This results in the generation of different vertical artifacts (B-lines) in the US image, which is directly related to the changes in the subpleural tissue. B-lines may indicate the accumulation of fluid in the pulmonary interstitial space or alveoli (Efremov et al. 2020). Multiple B-lines are associated with pulmonary edema of cardiogenic and noncardiogenic or mixed origin. These B-lines are generated when US waves transmit through the pleural line and encounters mixture of air and water. One or two B-lines are not too concerning; however, when they increase in number or spread out in one zone, they indicate lung interstitial syndrome and another key early marker of COVID-19 (Buonsenso et al. 2020). In recent studies, it has been confirmed that B-lines are heterogenic, and this can be used to distinguish the changes in the lung surface. Increase in the subpleural lung density can also be identified by multiple vertical artifacts in extended echogenic patters. In these cases, individual artifacts are sometimes resolved in the images or even may be lumped together to form a single echogenic area; this phenomenon is called "white lung" (Boppart and Richards-Kortum 2014).

3.6.2.3 Detection of COVID-19 Pneumonia

Figure 20 shows lung US images of a COVID-19 patient (52-year-old male) in which an irregular pleural line with subpleural consolidations, white lung area, confluence, and irregular artifactual B-lines are clearly visible (Buonsenso et al. 2020). Spared areas are also visible bilaterally, interspersed with pathological areas. In the same study, a control imaging experiment was performed on a COVID-19-negative patient where a normal pleural line with multiple reverberating A-lines was visualized in lung US. Also, only one artifactual B-line was noted in a single area.

Such case studies provide evidence that lung US is a valuable tool for assessing the pathological progression of COVID-19 pneumonia as shown in Fig. 21 (Smith et al. 2020). In summary, lung US imaging holds strong potential in the management of COVID-19 pneumonia in the ICU due to its safety, repeatability, low cost, and point-of-care use. With the increased use of bedside US in the ICU, patients can be protected from unnecessary radiation and therapy delays. The transport of high-risk patients to CT or X-ray examinations can also be avoided when PoC US is used.

3.6.3   Advanced Ultrasound Techniques with SARS Screening Potential

Although not very common in conventional clinical practice, a fusion of US imaging with other modalities like CT is essential for diagnosing complicated lung

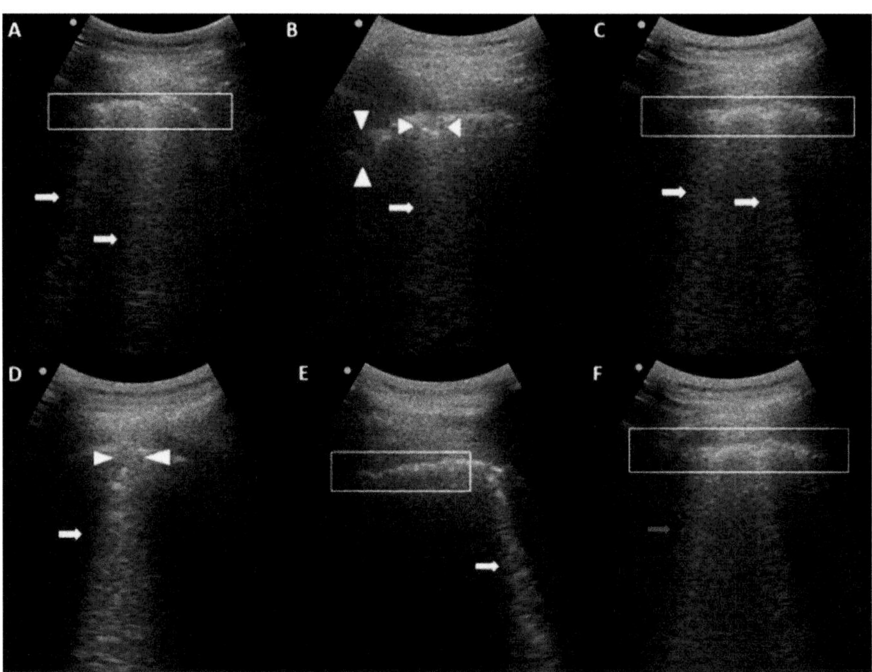

Fig. 20 Lung ultrasound findings in a patient with confirmed COVID-19 infection, showing pleural line irregularities (within the white boxes, figures **a-c-e-f**), thick irregular vertical artifacts (white arrows, figures **a-b-c-d-e**), subpleural consolidations (white arrowheads, figure **b-d**) and areas of white lung (red arrow, figure **f**). (Image reprinted from Buonsenso et al. 2020 with permission)

pathologies. This multimodal approach offers the advantages of both modalities by overlaying images with completely different contrast mechanisms. Recently, advanced US techniques like color Doppler, elastography, and contrast-enhanced ultrasound (CEUS) are also quite commonly exploited in clinical settings. For comprehensive screening and detailed characterization of lung vascularity, it would be ideal to combine these techniques with conventional pulse-echo imaging to get data-rich information from a single imaging session (Clevert et al. 2013). For example, Fig. 22 shows the fusion of conventional US imaging with Doppler to extract structural and blood flow details of a lung.

Photoacoustic (PA) imaging is an emerging imaging modality with advantages of both optical and US imaging. It beats the optical diffraction limit and offers resolution and imaging depth of US imaging, with signature optical contrast (Zhou et al. 2016). In PA imaging, short-pulsed light is illuminated on the tissue of interest; intrinsic tissue optical absorbers (hemoglobin, melanin, lipid, etc.) absorb light energy, resulting in thermoelastic expansion and subsequent generation of US signals. These US waves can be detected by using conventional US probes to generate tissue optical absorption maps. Since PA imaging involves US detection,

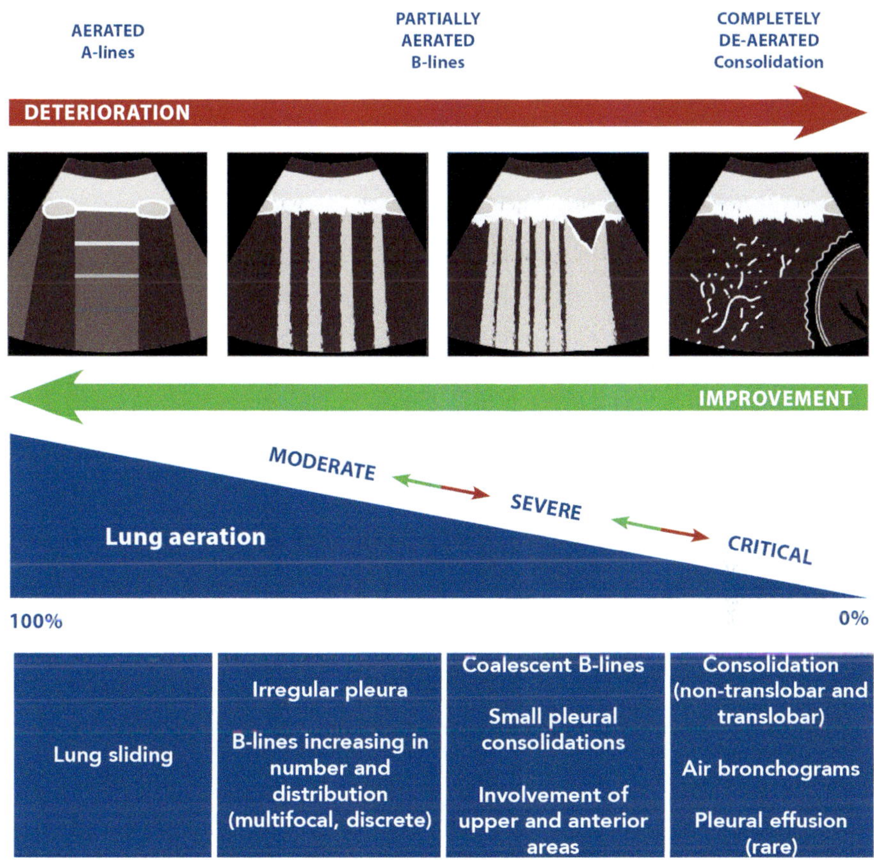

Fig. 21 Sonographic characteristics of moderate, severe, and critical pleural and parenchymal changes in COVID-19 patients. (Source: image reprinted from Smith et al. 2020)

it is seamless to develop portable and affordable dual-mode PA and US imaging systems with excellent structural and functional contrast by incorporating light delivery using portable light sources like LEDs (Zhu et al. 2020b; Singh et al. 2015). Since the optical absorption spectra of key tissue components are well known, it is possible to tune the light wavelengths and functionally characterize the tissue. For example, one can quantitatively detect oxygen saturation of a specific blood vessel with high temporal and spatial resolution, which is unachievable using any other PoC imaging techniques. When compared to US Doppler, PA imaging does not rely on flow, and one can get PA contrast even when there is no considerable blood flow (inflammation, clots, etc. can be imaged). Even though the use of PA imaging for lung screening is not well explored so far, it is foreseen that this promising technique may be suitable for assessing blood supply and lesion progression in peri-pulmonary consolidation at least in pediatric cases.

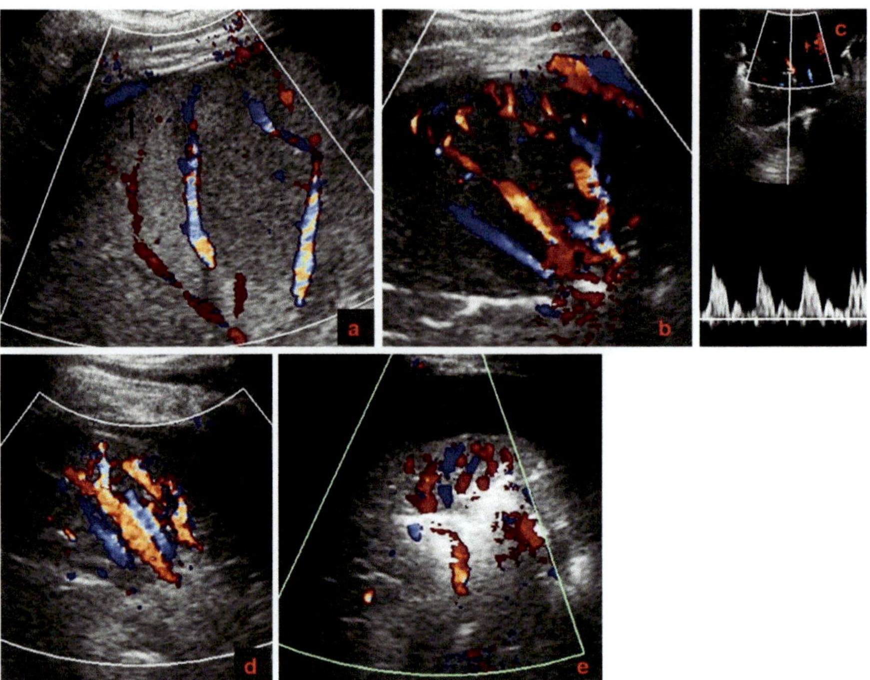

Fig. 22 Fusion of conventional ultrasound and color Doppler in the consolidated lung: (**a**) poorly vascularized pneumonia with no visible necrotic areas. The free movement of the pleural fluid produces color on Doppler US (arrow), (**b**) well-vascularized pneumonia with multiple normal branching vessels and, (**c**) quadriphasic wave pattern, and (**d–e**). different cases of atelectasis with multiple vessels having a crowded and parallel configuration, due to loss of the lung volume. (Image reprinted from Fygetaki et al. 2014 with permission)

4   Conclusion

Coronaviruses have evolved as dangerous human pathogen over the decades, primarily infecting the human pulmonary system causing moderate ARS to severe ARS. Classical medical imaging-based diagnostic procedures such as CT, X-rays, etc. provide high-resolution imaging-based assessment of the lung infection. These techniques are particularly useful for clinical decision-making in hospital settings. However, they are unsuitable for rapid screening and diagnosis of coronavirus disease especially during the COVID-19 pandemic. In addition to being time-consuming, costly, and nonportable, these imaging techniques also suffer from lack of unique and specific imaging markers that offer reliable diagnosis of the disease. Application of artificial intelligence (AI) for the identification of novel, unique, and reliable imaging markers may render these techniques useful during such pandemic situations.

Another well-known imaging technique that is immune to the common problems faced by CT and has a great potential for diagnostic use in case of coronavirus disease is ultrasound (US) imaging. US is one of the most popular radiation-free medical imaging modalities with unprecedented spatial resolution and imaging depth. The point-of-care US can provide superior details of the lung including fluid retention, without any radiation hazard when compared to conventional standard radiography, and, thus, generates quality information similar to X-ray or CT images. The gas present in the lungs creates huge acoustic impedance differences between lung tissue and the pleura, which covers the surface of the lungs. Therefore, a significant amount of transmitted US waves is reflected by the pleural layer resulting in a bright white structure of the lung surface on the image. However, when there is a change in the normal air-to-liquid ratio of lungs especially in conjunction with pleural disorders, there is a possibility of local US reflections resulting in unique pulse-echo contrast, which is of high clinical value. Although limited field of view offered by US imaging is a disadvantage compared to CT imaging, US imaging is inexpensive and particularly suitable for diagnostic purposes in resource-limited settings, where disease progression among medical professionals is also a serious concern.

Recent developments in miniaturized electronics and AI have resulted in the development and commercialization of wireless US imaging systems with mobile/tablet application support and the ability to rapidly analyze images. The computer-aided advancement facilitates the handling of the technology by untrained professionals to near perfection and thus democratizes the powerful medical imaging capabilities to regions and populations that may not have been able to benefit from the latest technological developments. AI adds diagnostic speed to the US capabilities in COVID-19 screening. The set algorithms, which had been trained on hundreds of thousands of US images (this is feasible considering the number of COVID patients around the globe), can enable the user to see patterns in the grayscale which allows the clinician to focus their attention on the task at hand. It can help doctors get a handle on the extent of an infection in a few seconds, allowing them to quickly make critical clinical decisions. Considering all these factors, the role of US imaging can be relevant in the context of coronavirus disease particularly during the COVID-19 epidemic.

Disclosures All the authors declare that they have no relevant financial interests or potential conflicts of interest to disclose.

References

Almazán F, González JM, Pénzes Z, Izeta A, Calvo E, Plana-Durán J, Enjuanes L (2000) Engineering the largest RNA virus genome as an infectious bacterial artificial chromosome. Proc Natl Acad Sci U S A 97(10):5516–5521

American College of Radiology (2020) ACR recommendations for the use of chest radiography and computed tomography (CT) for suspected COVID-19 infection. American College of Radiology. https://www.acr.org/Advocacy-and-Economics/ACR-Position-Statements/Recommendations-for-Chest-Radiography-and-CT-for-Suspected-COVID19-Infection

Bai H, Hsieh B, Xiong Z, Halsey K, Choi J, Tran T, Pan I, Shi L-B, Wang D-C, Mei J (2020) Performance of radiologists in differentiating COVID-19 from viral pneumonia on chest CT. Radiology:200823

Balasuriya UB, Barratt-Boyes S, Beer M, Bird B, Brownlie J, Coffey LL, Cullen JM, Delhon GA, Donis RO, Gardner I, Gilkerson J, Golde WT, Hartley C, Heidner H, Herden C (2017) Chapter 24 coronaviridae. In: Fenner's veterinary virology, 5th edn. Elsevier, pp 435–461

Bandyopadhyay S (2020) Coronavirus disease 2019 (COVID-19): we shall overcome. Clean Techn Environ Policy:1–2

Becker MM, Halpin RA, Li K, Venter E, Lu X, Scherbakova S, Graham RL, Baric RS, Stockwell TB, Spiro DJ, Denison MR, Eckerle LD (2010) Infidelity of SARS-CoV Nsp14-Exonuclease mutant virus replication is revealed by complete genome sequencing. PLoS Pathog:6–e1000896. https://doi.org/10.1371/journal.ppat.1000896

Bell D (2020) COVID-19. Radiology reference article. Radiopaedia.org. [Online]. Available: https://radiopaedia.org/articles/covid-19-3. Accessed 4 May 2020

Bernheim A, Mei X, Huang M, Yang Y, Fayad Z, Zhang N, Diao K, Lin B, Zhu X, Li K (2020) Chest CT findings in coronavirus disease-19 (COVID-19): relationship to duration of infection. Radiology:200463

Boppart SA, Richards-Kortum R (2014) Point-of-care and point-of-procedure optical imaging technologies for primary care and global health. Sci Transl Med 6:253

Brian DA, Baric RS (2005) Coronavirus genome structure and replication. In: Coronavirus replication and reverse genetics, Springer, pp 1–30

Buonsenso D, Buonsenso D, Raffaelli F, Piano A, Bonadia N, Franceschi F, Raffaelli F, de Gaetano Donati K (2020) Point-of-care lung ultrasound findings in novel coronavirus disease-19 pnemoniae: a case report and potential applications during COVID-19 outbreak. Eur Rev Med Pharmacol Sci 24(5):2776–2780

Bushmaker T, Morris D, Holbrook M, Gamble A, Williamson B, Tamin A, Harcourt J, Thornburg N, Gerber S, Lloyd-Smith J, de Wit E, Munster V (2020) Aerosol and Surface Stability of SARS-CoV-2 as compared with SARS-CoV-1. New Engl J Med. https://doi.org/10.1056/NEJMc2004973

CareMentor (2020) Care Mentor AI develops artificial intelligence for rapid diagnosis of COVID-19. CareMentor

Chopra D (2020) Teleradiology's role during COVID-19. M.C.I.C.E.O

Chung M, Bernheim A, Mei X, Zhang N, Huang M, Zeng X, Cui J, Xu W, Yang Y, Fayad Z (2020) CT imaging features of 2019 novel coronavirus (2019-nCoV). Radiology 295(1):202–207

Ciulla MM (2020) Coronavirus uses as binding site in humans angiotensin-converting enzyme 2 functional receptor that is involved in arterial blood pressure control and fibrotic response to damage and is a drug target in cardiovascular disease. Is this just a phylogenetic. J Med Virol

Clevert DA, D'Anastasi M, Jung EM (2013) Contrast-enhanced ultrasound and microcirculation: efficiency through dynamics-current developments. Clin Hemorheol Microcirc 53 (1–2):171–186

Cohen J, Normile D (2020) New SARS-like virus in China triggers alarm. Science 367 (6475):234–235

Deming D, Graham R, Denison M, Baric R (2007) Processing of open Reading frame 1a Replicase proteins nsp7 to nsp10 in murine hepatitis virus strain A59 replication. J Virol 81:10280–10291

Denison MR, Graham RL, Donaldson EF, Eckerle LD, Baric RS (2011) Coronaviruses: an RNA proofreading machine regulates replication fidelity and diversity. RNA Biol 8(2):270–279

Efremov SM, Kuzkov VV, Fot EV, Kirov MY, Ponomarev DN, Lakhin RE, Kokarev EA (2020) Lung ultrasonography and cardiac surgery: a narrative review. J Cardiothorac Vasc Anesth:30093–30098. https://doi.org/10.1053/j.jvca.2020.01.032

Escudero J, Ifeachor E, Zajicek J, Green C, Shearer J, Pearson S (2012) Machine learning-based method for personalized and cost-effective detection of Alzheimer's disease. IEEE Trans Biomed Eng 60(1):164–168

Faqin LV, Wang J, Yu X, Yang A, Liu J-B, Qian L, Xu H, Cui L, Xie M, Liu X, Peng C, Huang Y, Kou H, Wu S, Yang X, Tu B, Jia H, Meng Q (2020) Chinese expert consensus on critical care ultrasound applications at COVID-19 pandemic. Adv Ultrasound Diagn Ther 4(2):27–42

Farid A, Selim G, Awad H, Khater A (2020) A novel approach of CT images feature analysis and prediction to screen for Corona Virus Disease (COVID-19). Int J Sci Eng Res 11(3). https://doi.org/10.20944/preprints202003.0284.v1

Farid A, Selim G, Khater H (2020a) Applying artificial intelligence techniques for prediction of neurodegenerative disorders: a comparative case-study on clinical tests and neuroimaging tests with Alzheimer's Disease. Preprints, vol 2020030299. doi: https://doi.org/10.20944/preprints202003.0299.v1

Farid A, Selim G, Khater H (2020b) A composite hybrid feature selection learning-based optimization of genetic algorithm for breast cancer detection. Preprints. https://doi.org/10.20944/preprints202003.0298.v1

Farid A, Selim G, Khater H (2020c) Applying artificial intelligence techniques to improve clinical diagnosis of Alzheimer's disease. Preprints. doi: https://doi.org/10.20944/preprints202003.0297.v1

Fygetaki S, Tritou I, Stefanaki S, Antonopoulos S, Sfakianaki E, Heraklion G, Heraklion GR (2014) Pediatric chest: from x-ray to ultrasound. A pictorial review of 173 patients. Poster presented at ECR. https://doi.org/10.1594/ecr2014/C-0451

Ge X-Y, Li J-L, Xing-Lou Y, Chmura AA, Zhu G, Epstein JH, Mazet JK, Ben H, Zhang W, Cheng P, Zhang Y-J, Luo C-M, Bing T, Wang N, Yan Z, Crameri G, Zhang S-Y, Lin-Fa W, Daszak P (2013) Isolation and characterization of a bat SARS-like coronavirus that uses the ACE2 receptor. Nature 503:535–538

Goldman LW (2007) Principles of CT and CT technology. J Nucl Med Technol 35(3):115–128

Gordon DE, Jang G, Bouhaddou M, Xu J, Obernier K, White K, O'Meara M, Rezelj V, Guo JZ, Swaney DL, Tummino TA, Huettenhain R, Kaake RM, Richards AL, Tutuncuoglu B, Foussard H (2020) A SARS-CoV-2 protein interaction map reveals targets for drug repurposing. Nature:1–12

Gosert R, Kanjanahaluethai A, Egger D, Bienz K, Baker S (2002) RNA replication of mouse hepatitis virus takes place at double-membrane vesicles. J Virol 76:3697–3708

Hagemeijer MC, Rottier PJM, de Haan CA (2012) Biogenesis and dynamics of the coronavirus replicative structures. Viruses:3245–3269

Hu B, Zeng L, Yang X, Ge X, Zhang W, Li B, Xie J, Shen X, Zhang Y, Wang N, Luo D, Zheng X, Wang M, Daszak P, Wang L, Cui J, Shi Z (2017) Discovery of a rich gene pool of bat SARS-related coronaviruses provides new insights into the origin of SARS coronavirus. Plos Pathogens 13(11):e1006698

Jerome WG, Yoshimori T, Mizushima N, Denison MR, Prentice E (2004) Coronavirus replication complex formation utilizes components of cellular autophagy. J Biol Chem 279:10136–10141

King MQ (2012) Family coronaviridae. In: Virus taxonomy: ninth report of the international committee on taxonomy of viruses, Elsevier

Knoops K, Kikkert M, van den Worm S, Zevenhoven-Dobbe J, van der Meer Y, Koster A, Mommaas A, Snijder E (2008) SARS-coronavirus replication is supported by a Reticulovesicular network of modified endoplasmic reticulum. PLoS Biol 6.1957–1974

Kong W, Agarwal P (2020) Chest imaging appearance of COVID-19 infection. Radiol Cardiothoracic Imaging 2(1):e200028

Kooraki S, Hosseiny M, Myers L, Gholamrezanezhad A (2020) Coronavirus (COVID-19) outbreak: what the department of radiology should know. J Am Coll Radiol 17:447–451

Ksiazek T, Erdman D, Goldsmith C, Zaki S, Peret T, Emery S, Tong S, Urbani C, Comer JA, Lim W, Rollin P, Dowell S, Ling A, Humphrey C, Shieh W, Guarner J, Paddock C, Rota P, Fields B, DeRisi J, Yang J, Cox N, Hughes J, LeDuc JW (2003) A novel coronavirus associated with severe acute respiratory syndrome. N Engl J Med 348:1953–1966

Lai MM, Cavanagh D (1997) The molecular biology of coronaviruses. Adv Virus Res 48:1–100. Elsevier

Li W, Cui H, Li K, Fang Y, Li S (2020a) Chest computed tomography in children with COVID-19 respiratory infection. Pediatr Radiol:1–4

Li Q, Guan X, Wu P, Wang X, Zhou L, Tong Y, Ren R, Leung K, Lau E, Wong J, Xing X, Xiang N, Wu Y, Li C, Chen Q, Li D, Liu T, Zhao J, Liu M, Tu W, Chen C, Jin L, Yang R, Wang Q, Zhou S, Wang R, Liu H, Luo Y, Liu Y, Shao G, Li H, Tao Z, Yang Y, Deng Z, Liu B, Ma Z, Zhang Y, Shi G, Lam TTY, Wu JT, Gao GF, Cowling BJ, Yang B, Leung GM, Feng Z (2020b) Early transmission dynamics in Wuhan, China, of novel coronavirus-infected pneumonia. N Engl J Med 382(13):1199–1207

Luo C, Wang N, Yang X, Liu H, Zhang W, Li B, Hu B, Peng C, Geng Q, Zhu G, Li F, Shi Z (2018) Discovery of novel bat coronaviruses in South China that use the same receptor as Middle East respiratory syndrome coronavirus. J Virol 92(13):1–15

Mahapatra S, Chandra P (2020) Clinically practiced and commercially viable nanobio engineered analytical methods for COVID-19 diagnosis. Biosens Bioelectron 165:112361

McCall B (2020) COVID-19 and artificial intelligence: protecting health-care workers and curbing the spread. The Lancet Digital Health 2(4):e166–e167

Mossa-Basha M, Meltzer C, Kim D, Tuite M, Kolli K, Tan B (2020) Radiology department preparedness for COVID-19: radiology scientific expert panel. Radiology 200988. https://doi.org/10.1148/radiol.2020200988

Murray O, Bisset J, Gilligan P, Hannan M, Murray J (2020) Respirators and surgical facemasks for COVID-19: implications for MRI. Clin Radiol 75(6):405–407

Ng M-Y, Lee E, Yang J, Yang F, Li X, Wang H, Lui M, Lo C-Y, Leung B, Khong P-L (2020) Imaging profile of the COVID-19 infection: radiologic findings and literature review. Radiol Cardiothoracic Imaging 2(1):e200034

Oostra M, te Lintelo E, Deijs M, Verheije M, Rottier P, de Haan C (2007) Localization and membrane topology of coronavirus nonstructural protein 4: involvement of the early secretory pathway in replication. J Virol 81:12323–12336

Pan F, Ye T, Sun P, Gui S, Liang B, Li L, Zheng D, Wang J, Hesketh R, Yang L (2020) Time course of lung changes on chest CT during recovery from 2019 novel coronavirus (COVID-19) pneumonia. Radiology. https://doi.org/10.1148/radiol.2020200370

Rubin G, Ryerson C, Haramati L, Sverzellati N, Kanne J, Raoof S, Schluger N, Volpi A, Yim J-J, Martin I (2020) The role of chest imaging in patient management during the COVID-19 pandemic: a multinational consensus statement from the Fleischner society. Chest

Sahin AR, Erdogan A, Agaoglu PM, Dineri Y, Cakirci AY, Senel ME, Okyay RA, Tasdogan AM (2020) 2019 novel coronavirus (COVID-19) outbreak: a review of the current literature. EJMO 4(1):1–7

Salehi S, Abedi A, Balakrishnan S, Gholamrezanezhad A (2020) Coronavirus disease 2019 (COVID-19): a systematic review of imaging findings in 919 patients. Am J Roentgenol:1–7

Sazbo (2004) Diagnostic ultrasound imaging: inside out. New York, Elsevier Academic

Shan F, Gao Y, Wang J, Shi W, Shi N, Han M, Xue Z, Shen D, Shi Y (2020) Lung infection quantification of covid-19 in CT images with deep learning. arXiv:2003.04655

Simpson S, Kay F, Abbara S, Bhalla S, Chung J, Chung M, Henry T, Kanne J, Kligerman S, Ko J, Litt H (2020) Radiological society of North America expert consensus statement on reporting chest CT findings related to COVID-19. Endorsed by the Society of Thoracic Radiology, the American College of Radiology, and RSNA. Radiol Cardiothoracic Imaging 2(2):e200152

Singh MKA, Steenbergen W, Manohar S (2015) Handheld probe-based dual mode ultrasound/photoacoustics for biomedical imaging. In: Frontiers in biophotonics for translation medicine, Springer, pp 209–247

Smith MJ, Hayward SA, Innes SM, Miller ASC (2020) Point-of-care lung ultrasound in patients with COVID-19 – a narrative review. Anaesthesia. https://doi.org/10.1111/anae.15082

Soldati G, Smargiassi A, Inchingolo R, Buonsenso D, Perrone T, Briganti DF, Perlini S, Torri E, Mariani A, Mossolani EE, Tursi F, Federico M (2020) Is there a role for lung ultrasound during the COVID-19 pandemic? J Ultrasound Med. https://doi.org/10.1002/jum.15284

Staff H (2020) New imaging recommendations for suspected COVID-19. [Online]. Available: https://mobile.hospimedica.com/coronavirus/articles/294781251/new-imaging-recommendations-for-suspected-covid-19.html. Accessed 04 May 2020

Stephen NK, Gert UVZ, Louise N, Monique IA, Wolfgang P (2012) Human coronaviruses. In: Virology, Elsevier

Thackray LB, Miller BC, Lynn TM, Becker MM, Ward E, Mizushima NN, Denison MR, Virgin HW 4th, Zhao Z (2007) Coronavirus replication does not require the autophagy gene ATG5. Autophagy 3:581–585

van der Meer Y, Snijder E, Dobbe J, Schleich S, Denison M, Spaan W, Locker J (1999) Localization of mouse hepatitis virus nonstructural proteins and RNA synthesis indicates a role for late endosomes in viral replication. J Virol 73:7641–7657

Wang L, Wang Y, Ye D, Qingquan L (2020) Review of the 2019 novel coronavirus (SARS-CoV-2) based on current evidence. Int J Antimicrob Agents

World Health Organisation (2003) 31 12 2003. [Online]. Available: https://www.who.int/csr/sars/country/table2004_04_21/en/

World Health Organisation (2020) WHO Coronavirus Disease (COVID-19) dashboard, 11 5 2020. [Online]. Available: https://covid19.who.int/

Zhou Y, Yao J, Wang LV (2016) Tutorial on photoacoustic tomography. J Biomed Opt 21 (6):61007. https://doi.org/10.1117/1.JBO.21.6.061007

Zhu N, Zhang D, Wang W, Li X, Yang B, Song J, Zhao X, Huang B, Shi W, Lu R (2020a) A novel coronavirus from patients with pneumonia in China, 2019. N Engl J Med 382(8):727–733

Zhu Y, Feng T, Cheng Q, Wang X, Du S, Sato N, Yuan J, Singh MKA (2020b) Towards clinical translation of LED-based photoacoustic imaging: a review. Sensors 20:2484. https://doi.org/10.3390/s20092484

The Applications of Biosensing and Artificial Intelligence Technologies for Rapid Detection and Diagnosis of COVID-19 in Remote Setting

Syazana Abdullah Lim, Tiong Hoo Lim, and Afiqah Nabihah Ahmad

Abstract

COVID-19 is a new strain of coronavirus that had affected nations at a global scale. With an unprecedented high infection and mortality rate, the World Health Organization had declared this novel virus as a pandemic phenomenon in March 2020. Due to the seriousness of this situation, efforts to control and surveillance of this emerging disease are currently of global interest. This chapter will focus on analytical performance of biosensor and artificial intelligent (AI) technologies for the development of robust sensor to detect COVID-19. The future outlooks of biosensor and AI to be employed remotely for COVID-19 detection and diagnosis will also be discussed.

Keywords

Coronavirus · Biosensor · Artificial intelligence · Deep learning · Diagnostic tools · Remote settings

S. A. Lim (✉) · A. N. Ahmad
Center of Research for AgriFood Science and Technology (CrAFT), Universiti Teknologi Brunei, Bandar Seri Begawan, Brunei Darussalam
e-mail: syazana.lim@utb.edu.bn

T. H. Lim
Electrical and Electronic Engineering, Faculty of Engineering, Universiti Teknologi Brunei, Bandar Seri Begawan, Brunei Darussalam

P. Chandra, S. Roy (eds.), *Diagnostic Strategies for COVID-19 and other Coronaviruses*, Medical Virology: From Pathogenesis to Disease Control, https://doi.org/10.1007/978-981-15-6006-4_6

1 Introduction

Coronavirus (CoV) is a positive-strand virus with a crown-like appearance attributing to the existence of spiky glycoproteins on the envelope. Coronaviruses can be categorized into four genera: Alphacoronavirus (alphaCoV), Betacoronavirus (betaCoV), Deltacoronavirus (deltaCoV), and Gammacoronavirus (gammaCoV). It has been shown from characterization of genomes that alphaCoV and betaCoV have their gene originated from bats and rodents, while avian species are the gene sources of deltaCov and gammaCoV (Cascella et al. 2020). Coronavirus specifically affects the respiratory system or enteric tracts in animals and humans. The virus infects epithelial cells leading to otitis media, severe acute respiratory syndrome (SARS-CoV), and Middle East respiratory syndrome (MERS-CoV) (Kaya et al. 2020). For the past two decades, SARS-CoV, MERS-CoV, and severe acute respiratory syndrome coronavirus-2 (SARS-CoV-2) had caused three major outbreaks. Originating in China, SARS-CoV had triggered an epidemic affecting two dozen countries with an estimated of 8000 cases and 800 deaths. Middle East respiratory syndrome was first started in Saudi Arabia and had led to cases up to 2,500 with 800 deaths and is still being reported as isolated cases. Severe acute respiratory syndrome coronavirus-2 is the most recent outbreak caused by CoV that was named COVID-19 (an acronym of "coronavirus disease 2019") with its first detected case related to a seafood and wet animal market in Wuhan of Huibei Province in China (Cascella et al. 2020; Rothan and Byrareddy 2020). Although COVID-19 is similar to SARS-CoV, evidence revealed this virus displays epidemiological features that are markedly different from SARS-CoV (Heymann and Shindo 2020). Unlike SARS-CoV, individuals infected with COVID-19 generate a large amount of virus in the upper respiratory tract during a prodromal period, showed minimal symptoms and continue with their daily activities, which are the contributing factors for convenient transmission of this virus in a community. In contrast, containment of SARS-CoV outbreak was easier to achieve, as infection do not readily occur during prodromal period, and spread of infection to take place only during severe illness (Heymann and Shindo 2020). Due to the extraordinary and unprecedented scale of infection of COVID-19, this virus has been declared as pandemic by the World Health Organization in March 2020 where at the point of writing, over 2 million cases have been confirmed resulting in over 100,000 deaths (WHO 2020). At the present state, with no vaccine available, the only channel to manage COVID-19 disease involves a non-pharmaceutical intervention strategy of close monitoring of epidemiological incidences, isolation of COVID-19 patients, containment of positive cases, contact tracing, and vigorous surveillance system (Heymann and Shindo 2020; Mahapatra and Chandra 2020). As the spread of the disease remains uncontrollable, an efficient surveillance approach through robust testing approach is vital to cope with the transmission of COVID-19 among communities.

Real-time polymerase chain reaction (RT-PCR) assay has been used for the identification and detection of COVID-19 obtained from various clinical samples, such as oral swabs, throat swabs, and nasopharyngeal swabs, based on published lab protocols (Mao et al. 2020). As an alternative to RT-PCR, isothermal nucleic acid amplification technology has also been employed for fast detection of the disease, eliminating the needs of a series of alternating cycles of temperature where positive results can be obtained as short as 5 min and negative results up to 13 min (Abbott Laboratories 2020). However, these methods of identification and detection require reagents, expensive instruments, and highly skilled operators with molecular diagnostic expertise that are only accessible at a centralized laboratory. This makes real-time and effective monitoring of samples at isolated places with poor facilities difficult (Mao et al. 2020). Therefore, low-cost analytical tools and conveniently assessable systems that can quickly detect COVID-19 and isolate patients will make containment of the infectious virus—as part of a management crisis strategy—in a community highly feasible. Devices and tools to make up such systems are actively being researched on, and encompass various technologies, including biosensors, rapid assay, and real-time analysis tools, to name a few.

Precision medicine is a relatively new concept and being increasingly employed for diagnostic purposes by healthcare practitioners and researchers that allow effective formulation of disease prevention strategies by considering variability in a population, such as genes, and environment, as opposed to a one-fits-all concept. Precision medicine, combined with predictive medicine, a field to predict the probability of disease occurrence, will provide an invaluable tool in the management of COVID-19 pandemic crisis. Concurrently, the emerging fields of real-time analysis and rapid diagnostic tool in predictive and precision medicine have changed the traditional medical approach to infectious disease detection and containment. Data from different sources, such as patient health records, medical research and clinical trials, smart wearable biosensor and radiology devices, and social media, have been widely used for many medical applications, such as drug and medical device safety surveillance, healthcare service monitoring, and precision medicine (Reza and Najarian 2016). The adoption of high-quality computational-based data analysis has allowed medical practitioners and clinicians to understand the spread of COVID-19 and disease progression more efficiently. Computer-based systems can extract important clinical and statistical information related to the disease faster than human from the data sources. The applications of computer-based modelling and algorithms, such as machine learning (ML), artificial intelligence (AI), natural language processing (NLP), and deep learning, provide a faster and more precise data analysis of the virus outbreak and have been used for clinical decision support tools (Tuena et al. 2019). Machine learning algorithms operate by constructing models from real-time data inputs to make data driven predictions or decision making. They are commonly used in precision medicine. Precision medicine applies AI to analyze biosensor data to enhance diagnosis, prognosis, and the effectiveness of a particular treatment. The ML allows computers to build the data model and identify key hidden insights (Wynants et al. 2020). As more new data are collected, the computer will automatically learn and update the model to produce reliable, repeatable decisions and results.

This chapter will examine the ongoing advances of biosensors in developing diagnostic tools for COVID 19 pandemic surveillance monitoring in remote settings. Specific references to the experiences and technological advances of other CoVs will be made as an effort to circumvent the spread of virus. The different types of biosensor will not be discussed in detail and is beyond the scope of this chapter. Interested readers can refer to writings by Monosik et al. (2012), Takhistov (2005), Lim and Ahmed (2019), Zhao and Daubinger (2019), Rinken and Kivirand (2019), and Wu and Khan (2020) for an in-depth review of biosensors. The chapter is divided into two sections. The first part will focus on the biosensing technologies that have been developed for detecting CoVs, whereas the second section will describe how information that are obtained from biosensors can be utilized in COVID-19 detection and diagnostic, new drug discovery and repurposing, and outbreak monitoring through biological computational modelling and ML approaches.

1.1   Biosensor as a Diagnostic Tool

A biosensor is a device functioning as an analytical tool where it provides a signal corresponding to the amount of target analyte in a test sample. The signal output correlates proportionally to the concentration of target analyte in a pre-determined reaction (Lim and Ahmed 2019). In a typical configuration, a biosensor is made up of an analyte, a bioreceptor, and transducer as depicted in Fig. 1. Two examples of the commonly known and utilized biosensors are pregnancy detection kit and glucometer – a continuous glucose-monitoring biosensor for diabetic patients.

Figure 2 below provides a comprehensive overview of the classes of a biosensor. Biosensors can be categorized according to the types of bioreceptors and transducers employed. Biomolecular recognition is essential in biosensor application. In the beginning of the biosensor's development, recognition receptors were primarily obtained from living organisms. Examples of natural bioreceptors for biosensing applications are enzymes, whole cells, DNA and antibodies. With the emergence of modern techniques of recombinant technology, the engineering and manipulation of synthetic receptors in the laboratory have opened up endless possibilities for

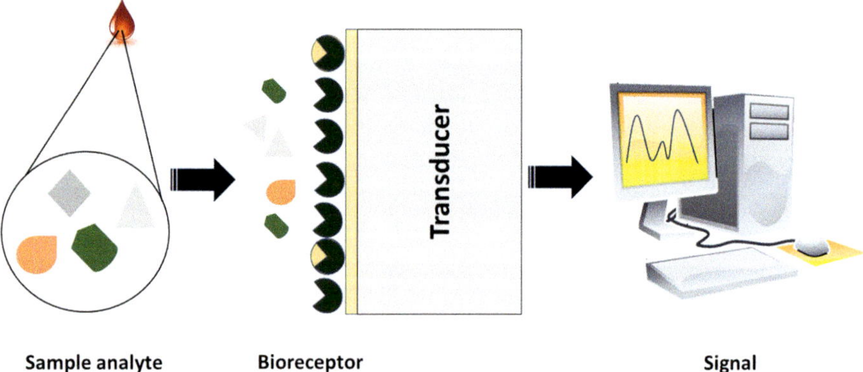

Fig. 1 Schematic diagram of a typical biosensor

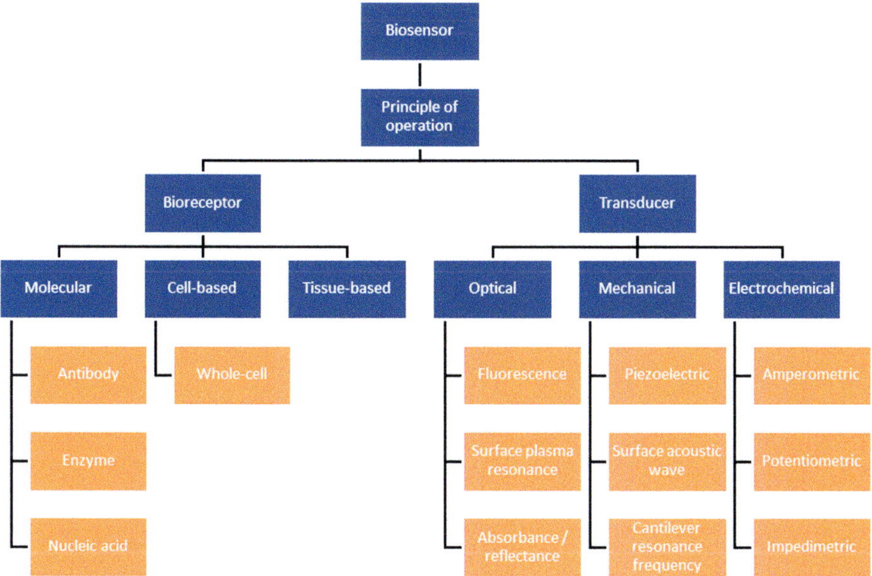

Fig. 2 Classification of a biosensor based on the bio-recognition receptors and transducers

biosensing design that is not possible undertaken by nature. Synthetic peptides and molecularly imprinted polymers are instances of such case. Depending on pre-determined biochemical reactions, biosensors can be further subdivided into the types of bioreceptors: biocatalytic and bioaffinity-based biosensors. A transducer is another important component of biosensors that works as an interface converting a biological event into a measurable signal. Based on the transduction mechanism applied, biosensors are categorized as electrochemical, optical, microgravimetric, and thermometric.

The usefulness of employing a biosensor as part of a COVID-19 crisis management strategy especially in places lacking in advanced facilities and laboratory can be attributed to its small size, which is highly practical for field work, and the small amount of sample. Furthermore, a biosensor possesses the advantages of fast response, highly specific and sensitive, and capable of real-time analysis without the need of skilled workers to operate required for testing (Lim and Ahmed 2019). These qualities render a biosensor as an ideal vehicle for point-of-care diagnostic tool with excellent potential for integration with latest technology for surveillance purposes in remote setting. For enhanced functions of a biosensor, various strategies have been put forward by researchers, such as improving the affinity between analyte and bioreceptors via novel surface chemistries and incorporation of nanomaterials in nanobiohybrid materials for signal amplification (Saylan et al. 2019; Lim and Ahmed 2016).

1.2 Intertwining Nanotechnology with Biosensing Strategies

Nanotechnology, among various fields of science and technology, holds a great promise in providing ground-breaking approaches to a wide range of problems related to the prevention, diagnosis, and treatment of COVID-19. The study in the field of nanomaterial-based biosensors is an interdisciplinary approach. It provides new opportunities for developing inexpensive and efficient detection methods, safe personal protection devices, and new effective medical strategies. The rapidly increasing death tolls of COVID-19 have been a wake-up call for global health. This sub-section will provide a brief overview with selected examples on the most recent advances of nanotechnology in fighting COVID-19 and other related viruses.

A nanomaterial has been defined as at least one of its dimensions acquiring less than 100 nm in size. Nanomaterials have been included in numerous biosensing strategies due to their extraordinary properties, which have shown excellent performance of biosensor development, in comparison to their bulk counterparts. In this approach, nanomaterials are often incorporated to amplify signal, for the ease of miniaturization and increase feasibility for technology integration. Depending on the types of detection techniques used, nanomaterials increase the sensitivity and overall performance of biosensors via various mechanisms. For instance, although there is abundance of excellent characteristics acquired by nanomaterials, only two of their attributes are of importance in developing electrochemical biosensors—high surface area and outstanding electrical conductivity. Signals of electrochemical output are enhanced by nanomaterials by reducing detection potentials, escalating current output, enhancing stability and resistance to electrode fouling, and bettering biomolecule compatibility and functionalization (Lim and Ahmed 2016). For fluorescence-based detection, such as fluorescence resonance energy transfer (FRET), this sensing format can benefit from the large surface area of nanomaterials and their unique-sized optical properties for improved signal yield. Plasmonic nanomaterials normally function as nano-quencher (energy acceptor) and nano-scaffold. Due to their nano-sized dimension (<10 nm), the effect of quantum confinement leads to large UV-visible absorption spectra, narrow emission bands, and optical attributes that can be manipulated through dimension, structure, and shape.

Biofunctionalization of nanomaterials has been the current trend in the development of nanobiosensors (Lim and Ahmed 2016). Through the formation of bioconjugate nanocomplexes, these interesting materials play important roles as catalysts to facilitate the electroactive species yield (nanocatalysts), redox active species (nanorecorders), and vehicles for reporter molecules (nanocarriers) (Bezinge et al. 2020). Diverse forms of nanomaterials have been employed in virus detections, such as nanorods for Hepatitis B virus (Draz and Shafiee 2018), nanoparticles for Zika and dengue viruses (Steinmetz et al. 2019; Chen et al. 2009), nanowires for Ebola virus (Generalov et al. 2019), nanotubes for Influenza A virus (Tam et al. 2009), and many others have been successfully applied to develop numerous biosensors. Nano-bioconjugates have been used to increase the sensitivity, precision, and efficiency of biosensors for the detection of CoVs. As an illustration, zirconium quantum dots (Zr QDs), a nanocrystal, were synthesized that employed L (+)

ascorbic acid to serve as a surface and chiral ligand for the determination of infectious bronchitis virus (IBV) of coronavirus (Ahmed et al. 2018). In this work, the fluorescence-based immunosensor was fabricated by conjugations of antibodies to Zr QDs and magneto-plasmonic nanoparticles producing a magnetoplasmonic-fluorescent nanohybrid arrangement to form an immunolink in the presence of the IBV. An external magnet was then applied to separate the nanostructured hybrid to determine the concentration of target analyte by measuring the photoluminescence intensity of magnetoplasmonic-fluorescent nanohybrid. An increase in signal intensity by 25% was reported in nanohybrids bioconjugates in comparison with antibody-conjugated Zr QDs. The detection scheme is represented in Fig. 3. A limit of detection (LOD) of 79.15 EID/50 μL was obtained for detection of the CoV.

Recently, Qiu et al. (2020) developed dual-functional plasmonic photothermal biosensors using two-dimensional gold nanoislands functionalized with complementary DNA receptors that were able to perform a sensitive detection of the selected sequences from COVID-19 through nucleic acid hybridization. The invented biosensor has shown to exhibit a high sensitivity towards the selected SARS-CoV-2 sequences with lower detection limit down to the concentration of 0.22 pM, and this allows precise detection of the specific target in a multigene mixture (Qiu et al. 2020). Additionally, Kim et al. fabricated a label-free colorimetric assay using self-assembled gold nanoparticles (AuNPs) to detect MERS-CoV. Colorimetric biosensors are extremely useful in cases of outbreak, owing to their low-cost, simplicity, portability, and disposability. This technique primarily determines the presence of target analytes through change of color of substrate with the aid of color reagents. The designed platform took advantage of the excellent optical property of AuNPs that was able to detect the target virus through a localized surface plasmon resonance (LSPR) shift and color changes of AuNPs in the UV−vis wavelength range. In this scheme, self-induced disulfide bond formation from the hybridization of complementary target dsDNA and target had prevented AuNPs aggregation, causing prominent color and wider red-shift of the LSPR peak (Fig. 4). Furthermore, this colorimetric assay could discriminate down to 1 pmol/μL of 30 bp MERS-CoV, and due to its convenience, it could be tailored for on-site detection of other infectious diseases, particularly in resource-limited settings.

Ishikawa et al. (2009) had developed nanowires biosensors utilizing antibody mimic proteins as probes to detect SARS. It was concluded that the sensitivity of biosensors can be obtained in a relatively short time and without the aid of any signal amplifier, such as fluorescence-labelled reagents. This demonstrates the potential for nanobiosensors to be used as an accurate, convenient, and rapid tool to detect biological complex systems. With the discovery of novel diseases, the role of nanotechnology especially in medicine has been the most emerging and highly researched science where the role of biosensors is drastically sought for. Traditional biosensors possess few bottlenecks, namely sensitivity and selectivity. These drawbacks can be potentially overcome if transducer system in the biosensor is based on nanomaterials. Nanoscale particles increase the sensitivity, speed, and flexibility of biological tests measuring the presence or activity of selected substances. Moreover, it enables testing of relatively small sample volumes. The

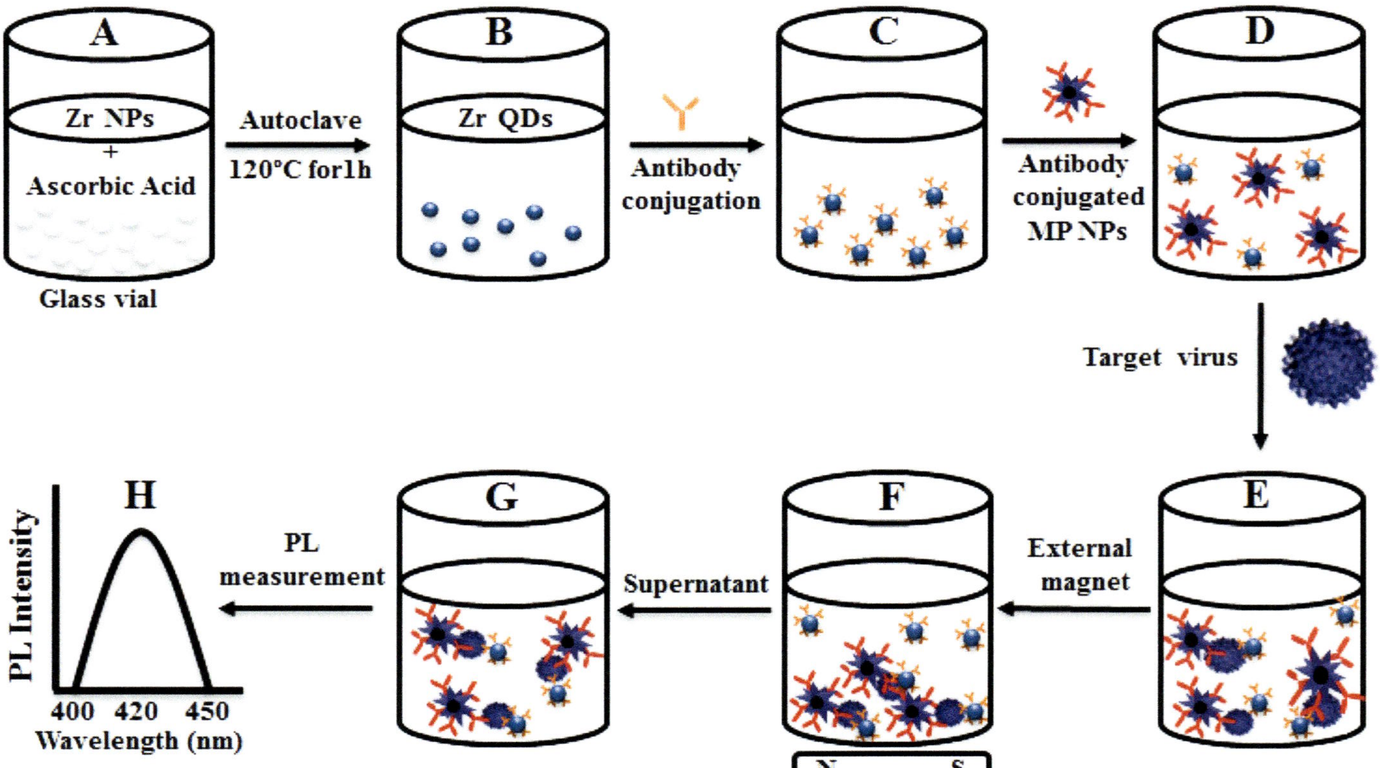

Fig. 3 Schematic illustration of sensor to detect infectious bronchitis virus of coronavirus: (**a**) Zr NPs and ascorbic acid as reducing agent; (**b**) Zr QDs formed; (**c**) bioconjugate-QDs; (**d**) the addition of antibody-conjugated MP NPs; (**e**) production of nanostructured magnetoplasmonic-fluorescent, target was then added and later separated (**f**); (**g**) the nanohybrid-conjugated part was separated, and the resulting optical properties were determined (**h**). (Copyright, Elsevier, Ahmed et al. 2018)

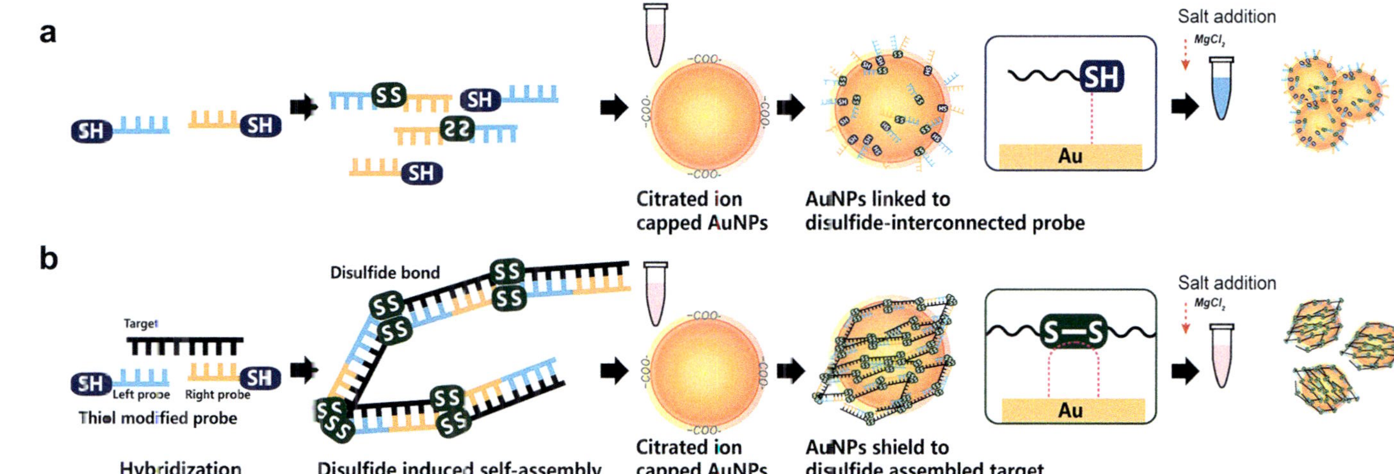

Fig. 4 Colorimetric detection scheme of MERS-CoV DNA, using disulfide-induced self-assembly strategy. (**a**) Salt-induced aggregation of AuNPs without the presence of target analyte. (**b**) Steps involved to inhibit AuNPs from salt-induced aggregation by disulfide-induced self-assembly in the presence of targets. (Copyright, ACS Sensors, Kim et al. 2019)

advancement of nanotechnology has undoubtedly provided nanofabricated devices that are small, sensitive, and inexpensive enough to facilitate direct observation and analysis. In the near future, nano-diagnostics would reduce the waiting time for the test results.

1.3 Recent Strategies for Rapid Coronavirus Detection

1.3.1 Immunosensor-Based Detection

Antibodies, which can be polyclonal or monoclonal, are regarded to be the prime choice used in biomolecular recognition component in biosensor due to their target specificity and affinity. In a recent work by Seo and co-workers, the group designed a graphene-field-effect transistor (FET)-based biosensing device as a diagnostic instrument for the detection of COVID-19 (Seo et al. 2020). Figure 5 displays a schematic representation of their biosensor setup. Spike protein of COVID-19, a major transmembrane protein of the virus and has shown high immunogenicity, was employed as a biomarker and diagnostic antigen. COVID-19 spike antibody was immobilized onto the COVID-19 FET sensor through a probe linker, 1-pyrenebutyric acid N-hydroxysuccinimide ester. The real-time performance of their sensor was assessed by evaluating the LOD for spike COVID-19 protein in PBS and successfully achieved an impressive LOD of 1 fg/mL (Fig. 6). Furthermore, the researchers tested their COVID-19 FET immunosensor using real clinical samples of nasopharyngeal swabs suspended in universal transport medium with the device detected the protein when diluted as much as $1:1 \times 10^5$ (242 copies/mL) in less than 1 min. Although this technique was found to be 2–4 times less sensitive

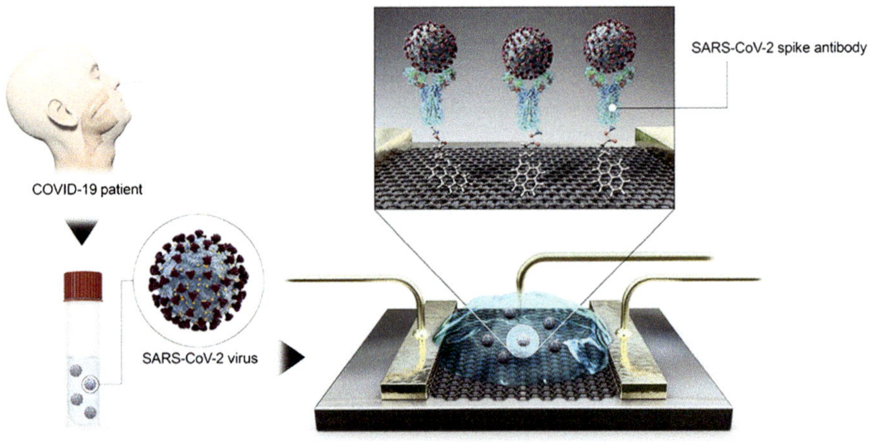

Fig. 5 Detection principle of COVID-19 FET sensor operation procedure. Graphene sheet was employed, SARS-CoV-2 spike antibody is linked with the graphene sheet via a probe linker of 1-pyrenebutyric acid N-hydroxysuccinimide ester. (Copyright, ACS Publication, Seo et al. 2020, https://doi.org/10.1021/acsnano.0c02823)

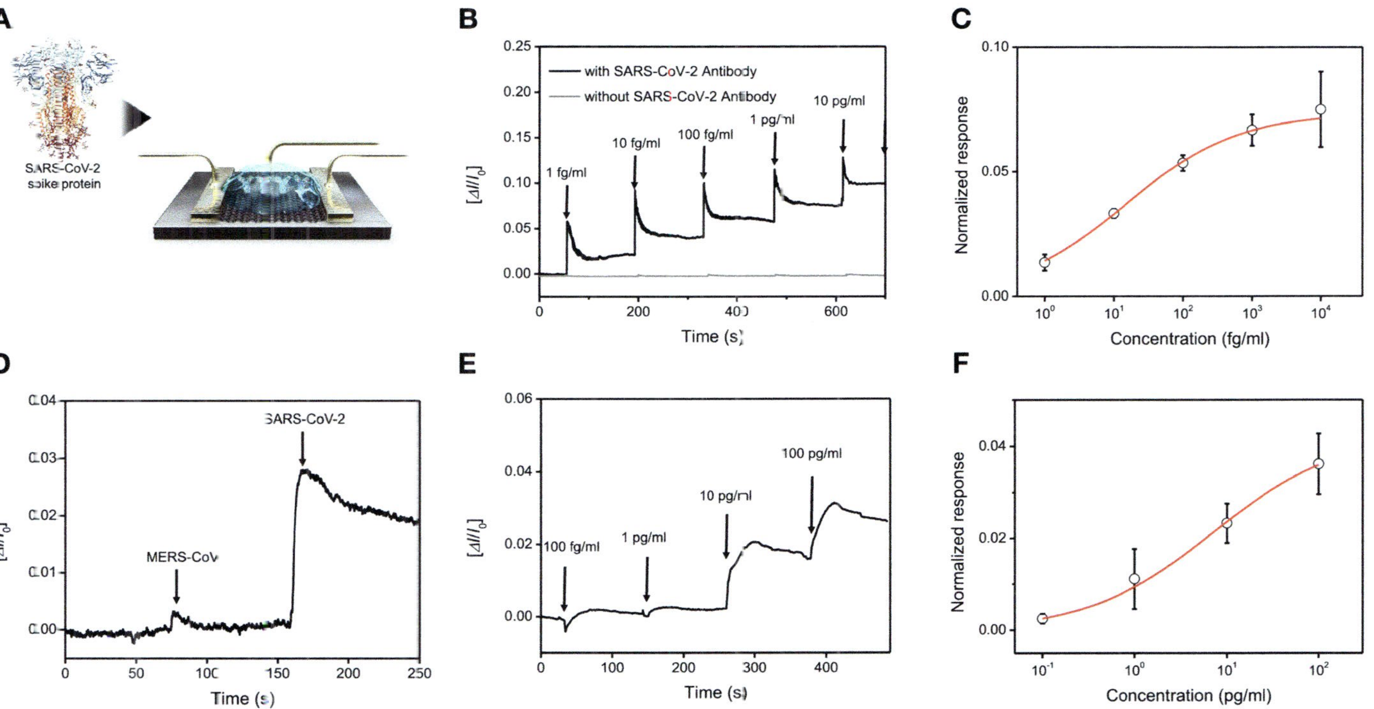

Fig. 6 Determination of SARS-CoV-2 antigen protein. (**a**) Principal of detection to determine the presence of SARS-CoV-2 spike protein. (**b**) Real-time response of COVID-19 FET towards SARS-CoV-2 antigen protein in PBS. (**c**) Related dose-dependent response curve ($V_{DS} = 0.01$ V). Graphene-based FET as negative control in the absence of SARS-CoV-2 antibody. (**d**) Selectivity of COVID-19 FET sensor towards target SARS-CoV-2 antigen protein and MERS-CoV protein. (**e**) Real-time result of COVID-19 FET towards SARS-CoV-2 antigen protein in UTM. (**f**) Related dose-dependent response curve. (Copyright, ACS Publication, Seo et al. 2020, https://doi.org/10.1021/acsnano.0c02823)

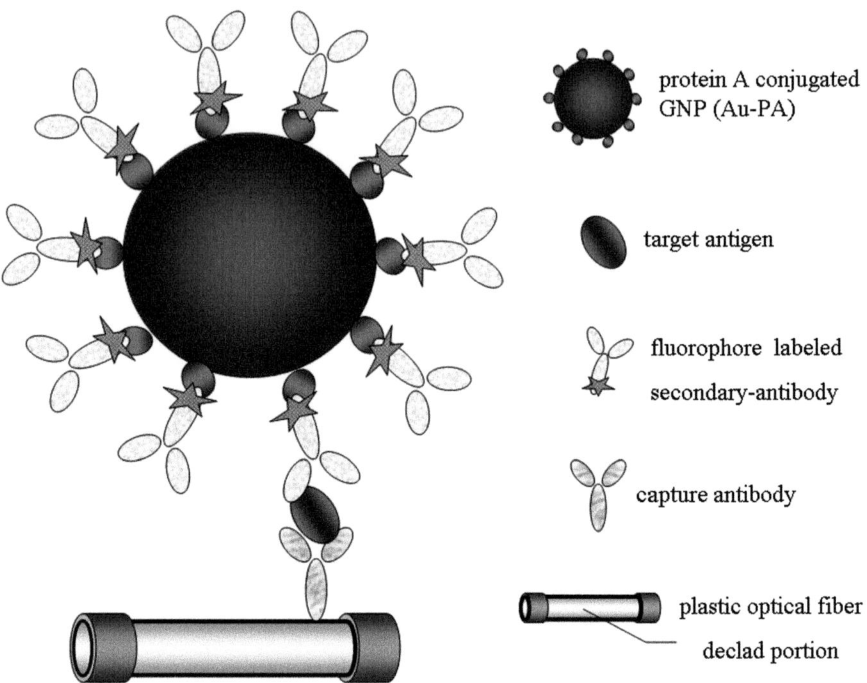

Fig. 7 Procedure for the formation of sandwich complex and fluorescence probe. (Copyright, Elsevier, Huang et al. 2009)

than the current molecular diagnostic tests (LOD of the virus is ~50–100 copies), the background noise still could be further improved by exploring other materials for detection platform. Field-effect transistor-based sensors are deemed to be potentially resourceful for clinical diagnosis and on-site applications.

Nucleocapsid (N) protein is also another valuable biomarker for early and accurate diagnostic approach, since this protein is expressed early and can be detected as early as 1 day upon infection of SARS. An optical-based LSPR device to detect N protein was designed for SARS-CoV based on a sandwich immunosensor format utilizing fluorescence probe as signal output (Huang et al. 2009). The fundamental principle behind optical biosensor's ability to detect analyte relies on the fact that all proteins, cell, and DNA have dielectric permittivity larger than air and water that caused these biological molecules to slow the propagation speed of electromagnetic fields passing through them. Optical biosensors quantify any alteration in phase, speed of polarization, or frequency of input light upon recognition of bioreceptors by their targets passing through them (Cooper 2009). Antigen recombinant SARS-CoV N protein (GST-N protein) was synthesized for target. Anti-N-1 and anti-N-2 antibodies were used as bioreceptors. A fluorophore was labelled with protein A conjugated with AuNPs and were attached to the secondary antibodies as shown in Fig. 7. Their LSPR fiber-optic biosensor achieved an LOD of 0.1 pg/mL. In addition, they tested their method in ten-fold diluted human serum and could detect the target antigen at 0.1 pg/mL, which is equivalent to 1 pg/mL in undiluted serum. The group attributed their enhanced sensitivity to several factors: fluorescence was enhanced near the gold nanoparticles surface; each fluorescence probe was able to accommodate many fluorophores on its surface; and the protein A served as a spacer that had prevented the quenching effects of metal. Nanobiosensors have shown great ability to detect bacteria and viruses at very low concentrations and thus, own the capability to warn clinicians even before symptoms appeared or on patients with very low viral loads.

Microfluidic involves the manipulation of small volumes of fluids, as a way to govern the state of chemical, biological, and physical activities with regards to sensing (Stroock 2008). A microfluidic immunosensor based on cotton thread has been developed for the fabrication of cheap and low-volume diagnostic device for rapid detection of infectious bronchitis virus (IBV), an avian coronavirus (Weng and Neethirajan 2018). Fluorescence resonance energy transfer was employed as the sensing mechanism, and molybdenum disulfide (MoS_2), a two-dimensional nanosheet, was used as a fluorescence quencher on fluorophore when antibody–antigen complex was formed. In their detection scheme, fluorescent dye labels and MoS_2 were attached to two antibody probes, and these bioconjugates form immunocomplexes in the presence of target IBV. Upon formation of immunocomplexes, the fluorescence of the fluorescent-labelled conjugate probe is mostly quenched. The LOD was calculated to be 4.6×102 EID_{50} per mL.

1.3.2 DNA-Based Biosensor

In a biosensor where DNA is employed as a bio-recognition element, a single-stranded RNA is attached on a transducer surface to detect its complementary DNA sequence. The consequent surface hybridization leads to the formation of hybrid, and this event will then be translated into an analytical signal by a transducer. A LSPR-DNA-based biosensor was developed by Jing Wang et al. as a practical alternative to the current PCR diagnostic tool for the identification of COVID-19 virus. AuNPs were functionalized with complementary DNA receptor probes for detection of COVID-19 two specific sequences, the RdRp and the ORF1ab, through nucleic acid hybridization. A thermoplasmonic heat was used to heat the gold nanoparticles to increase the in situ hybridization temperature and improve the discrimination of two similar COVID-19 gene sequences thus increasing the selectivity of this method against SARS-CoV and reducing the possibilities of false-positive results. This biosensor reported a high sensitivity and specificity towards COVID-19 RNA gene sequences and a LOD of 0.22 pM. Its actual potential use for real clinical applications is yet to be implemented and evaluated especially in the presence of multiple nonspecific sequences in a real sample.

Simultaneous detection of multiple analytes in a sample is a valuable clinical diagnostic tool. Shi et al. (2015) developed an optical nucleic acid-based biosensor, with surface plasmon resonance (SPR) as the transducer, aimed at identifying multiple respiratory viruses, including SARS in a single detection platform. Specific RNAs were labelled with biotin and immobilized on a SPR chip where streptavidin

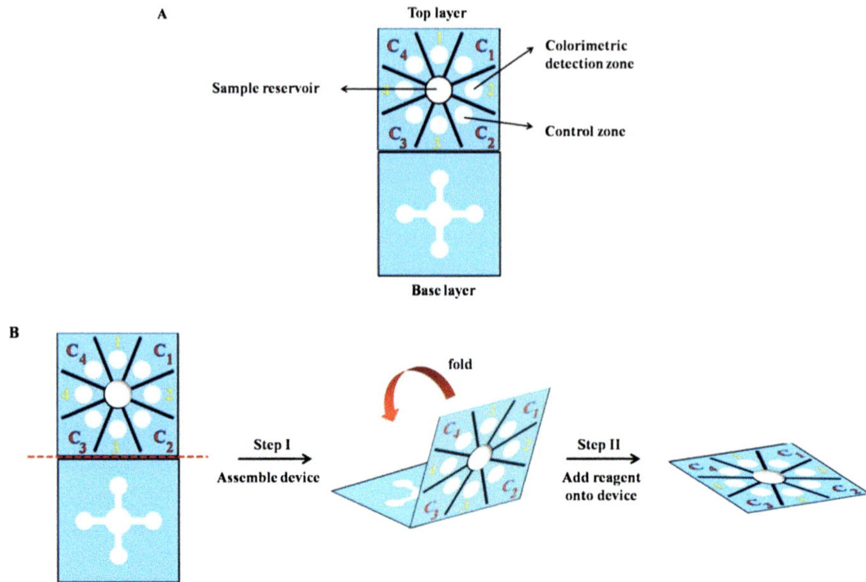

Fig. 8 (**a**) Design and (**b**) Work procedure of Multiplex Paper-Based Colorimetric Device. (Copyright, ACS Publication, Teengam et al. 2017, https://doi.org/10.1021/acs.analchem.7b00255)

was later introduced post hybridization for signal amplification. The group reported a LOD of 2 nM with excellent repeatability.

A proof-of-concept DNA-based colorimetric assay was developed based on pyrrolidinyl peptide nucleic acid (acpcPNA)-induced silver nanoparticle (AgNP) aggregation as an alternative to the typical colorimetric approaches using AuNPs (Teengam et al. 2017). The multiplex colorimetric sensor utilized paper substrate as their platform to determine MERS-CoV, Mycobacterium tuberculosis (MTB), and human papillomavirus (HPV). The positively charged acpcPNA was designed specifically to perform as a probe that interacted with the negative AgNPs, which caused aggregation of nanoparticle and subsequent color change. As shown in Fig. 8, the top part of the paper-based configuration consisted of four zones for detection and control, respectively. To ensure DNA selectivity, each zone contained AgNPs with only single acpcPNA. The layer at the base constituted four wax-defined canals that extended outward from an opening to receive sample. These base channels were linked to four detection zones of the top layer when the paper platform was folded and stacked. Color change could be observed upon addition of target sample via the outward movement of solution through the base channels to the zone where detection took place as depicted in Fig. 9. It was reported that MERS-CoV could be detected within a linearity range of 20–1000 nM and LOD of 1.53 nM.

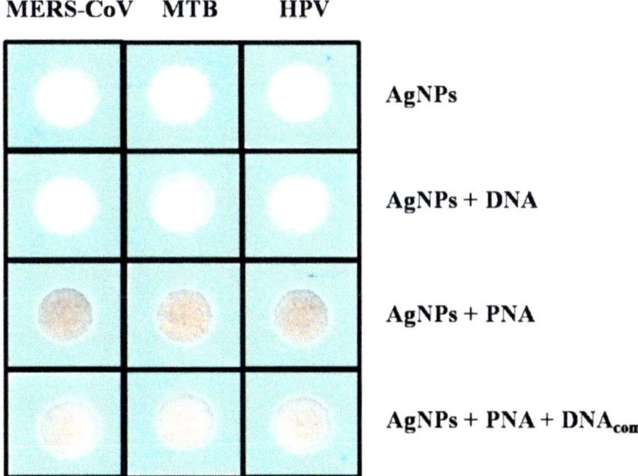

Fig. 9 Picture to demonstrate visual color changes based on detection of MERS-CoV, MTB, and HPV in the presence of DNA_{com}. (Copyright, ACS Publication, Teengam et al. 2017, https://doi.org/10.1021/acs.analchem.7b00255)

2 Non-contact Diagnostic Technologies

Conventional analysis of clinical and laboratory data presents a number of challenges on the prediction of disease progression and diagnosis. The processes of understanding the properties and predicting the behavior of COVID-19 or other CoVs are not easy and fast enough to provide a proper treatment and diagnosis, and also drug and vaccine development as most clinical datasets are multimodal and have missing observations. Dealing with these issues typically requires extensive pre-processing that can be achieved using computer-based ML. For example, Esteban et al. (2017) focused on four data variables from 200,000 intensive care unit patient records to predict the COVID-19 disease progression. Developing algorithms and methods that can overcome these limitations is a key step towards broader applications of computer-based ML in precision medicine.

2.1 Computer-Based Data Modelling for Remote Diagnostic and Detection

Computer-based modelling has been widely used to precision medicine for remote disease detection and diagnosis using real-time data (Ting et al. 2020). To process and analyze clinical or real-time data obtained from the biosensor, ML process as shown in Fig. 10 can be applied to provide remote diagnosis to tackle COVID-19 pandemic based on the information obtained from the biosensor. Patients who are positive or at risk for COVID-19 can be attached with a biosensor on their arms and sent home for remote monitoring. The sensors can collect different physiological signals, such as temperature, cough activities, oxygen blood level, and heart rate from the wearable and compared with clinical data in the healthcare database.

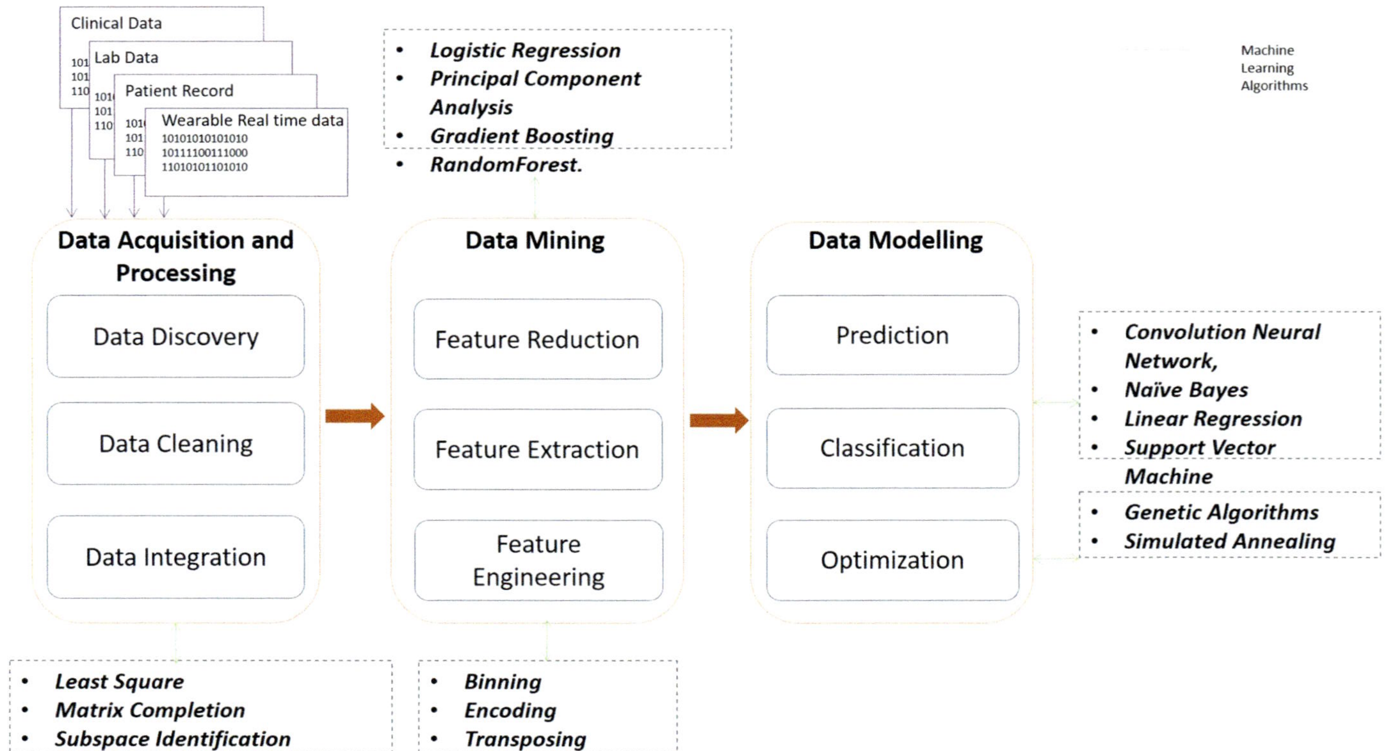

Fig. 10 Machine Learning Process for Medical Diagnostics and Detection

The development of a remote medical diagnostics and detection system consists of three parts namely data pre-processing, data mining, and data modelling as shown in Fig. 10. In order to remove anomalies and missing data recovery, the sensor and medical data collected from embedded devices need to be processed and cleaned during the data pre-processing. Some of the algorithms commonly used in data pre-processing are least square, linear regression, and subspace identification. For instance, missing data can be inserted using linear regression, and data anomalies can be remove using nearest neighbor approaches during COVID-19 (Pirouz et al. 2020).

Once the data pre-processing is complete, important features related to symptoms of the COVID-19 can be extracted using data mining method, such as Sequential Forward Selection, Sequential Backward Selection, and Recursive Feature Elimination before they are fed into an AI algorithm for data analysis and model generation (Fahmiin and Lim 2020). The AI algorithm will generate a model that can mimic the behavior of the COVID-19 and used to detect the CoV using the diagnostic information as data input.

As more data are available, the model can be optimized using parameter-tuning approach to improve the physiological signature of the virus in order to provide early detection and effective treatment.

2.2 Artificial Intelligence in Biomedical Engineering

High-performance cloud technologies have made significant contributions especially in detecting, diagnosing, and controlling the spread of the COVID-19 pandemic (Senior et al. 2020). Such technologies are used for image processing, data analytics, Internet of Things (IoT), and AI (Allam and Jones 2020). Artificial intelligence has not only helped in early diagnosis and development of the disease through the deployment of biosensor, it is also used in the area of computational protein and molecular modelling for identifying the best treatment and drug development, biomarker analysis, and for pandemic tracing as shown in Fig. 11 (Chen et al. 2020).

Although AI systems usually require high-performance computer running on clouds to process the large amount of data, medium range-embedded computers have also been used to perform early diagnostics and provide secondary care to users. For example, intelligent robots and drones are used in hospitals and at home to deliver food and medicine supplies and testing kits during lockdown. Many university academics and researchers have developed medical devices and produce drugs and vaccine to treat and prevent the virus using AI simulation model (Stebbing et al. 2020). Computational algorithms have been applied to detect infectious COVID-19 patients using medical imaging data, such as X-ray images and Computed Tomography (CT) scans, cough samples, temperature scan, and blood and swab test samples from biosensor (Shi et al. 2020).

This section will discuss the three common areas where AI has been applied in COVID-19, and these are:

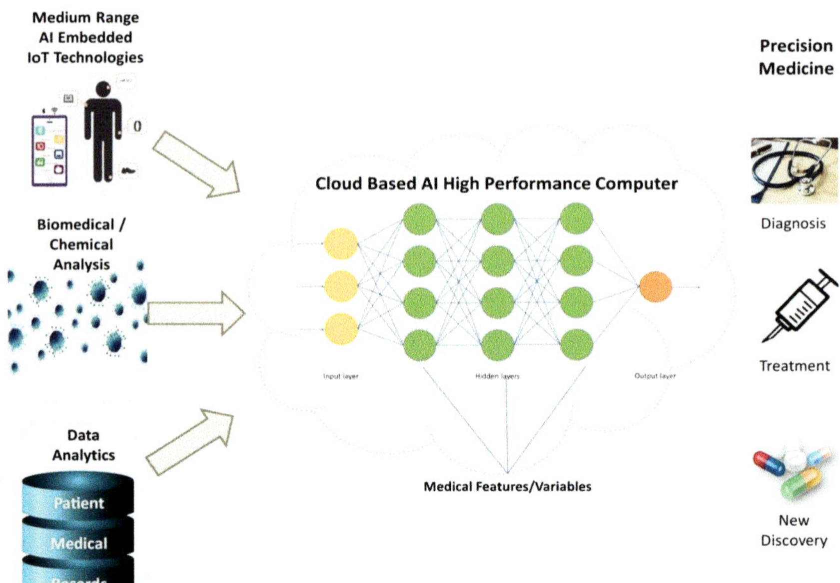

Fig. 11 Cloud-based Artificial Intelligence in Precision Medicine

1. *Computational biological modelling for biosensor and biomarker*
2. *Fast Diagnosis using deep learning methods*
3. *Pandemic tracking using sensor data analytic*

2.2.1 Computational Biological Modelling for COVID-19 Treatment

With the number of infections that had increased dramatically, it is important to understand the biochemistry properties of the proteins collected from the biosensor. To investigate the functions of the proteins related to COVID-19 virus, Senior et al. have applied the template-free modelling on the Google Deepmind's AlphaFold cloud systems to predict the structures and roles of proteins related to COVID-19, including the membrane protein, protein 3a, nsp2, nsp4, nsp6, and papain-like protease (Senior et al. 2019; Jumper et al. 2020). Senior et al. applied the neural network to make accurate predictions of the distances between pairs of residues that provide more information about the structure. They create a potential of mean force to accurately describe the shape of a protein and optimized the result using a simple gradient descent algorithm to generate structures without complex sampling procedures using the processes as illustrated in Fig. 12.

Google cloud has been widely used for protein structure predictions using deep learning (Wei 2019). The COVID-19 proteins have a 3D structure that is determined by the genetically encoded amino acid sequence. Four types of structural proteins from COVID-19 of interest are nucleocapsid proteins, envelope proteins, membrane proteins, and spike protein. Although the authors have released the predicted

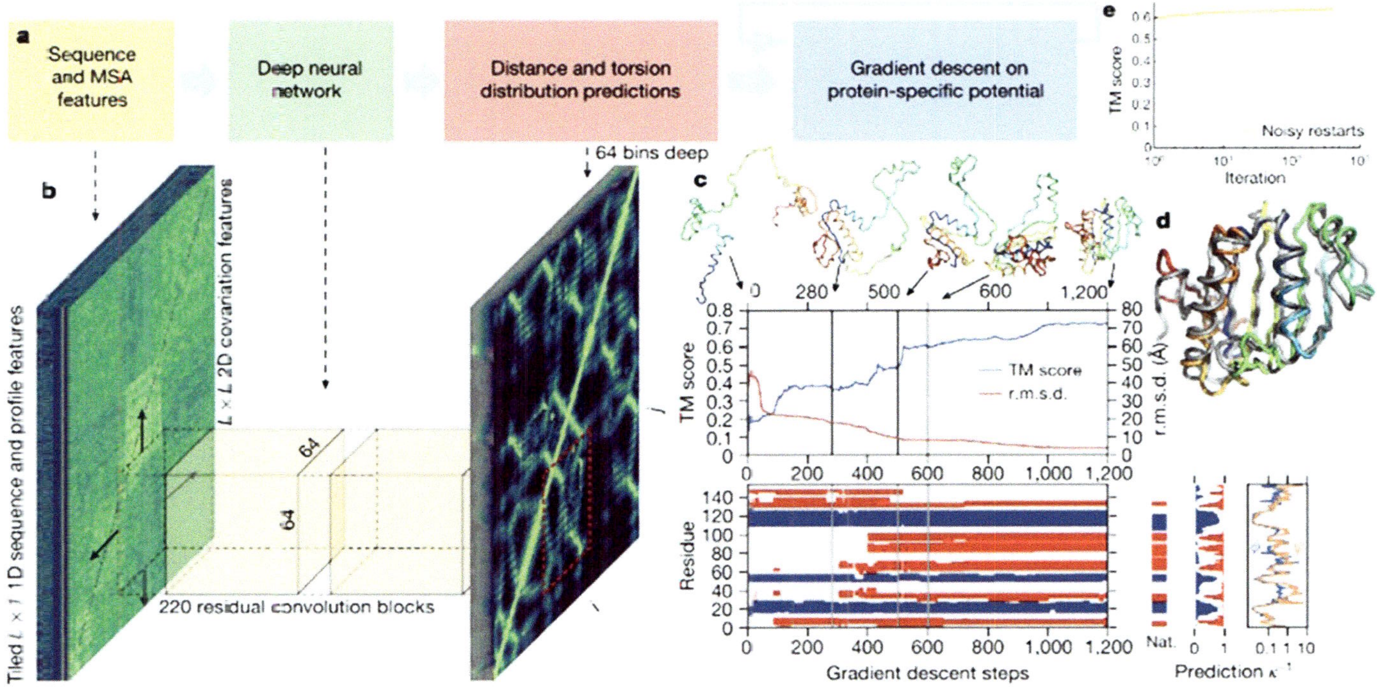

Fig. 12 The flooding process of the Alphafold systems to predict the structure from protein sequence: (**a**) Steps of structure prediction; (**b**) The neural network predicts the entire L x L distogram based on MSA features; (**c**) One iteration of gradient descent with the TM score and RMS deviation plotted against step number; (**d**) The final native structure; (**e**) The average TM score of the lowest-potential structure against the number of gradient descents per target. (Copyright, Springer Nature, Senior et al. 2020)

structures, these structures still need to be validated as there is no experimental evidence to support any of the conclusions.

To address the lack of experimental evidence to support several proteins, Zhang et al. (2020) have used the C-I-TASSER15 to create structural models of the full-length spike protein. The C-I-TASSER15 is an extension to I-TASSER15 using Convolutional Neural Network to guide the Monte Carlo fragment assembly simulation to reassemble the domains and to construct a complex structure consisting of the spike trimer and the extracellular domain of human ACE2. The authors have found out that the viral sequence from intermediate hosts of SARS-CoV and MERS-CoV are 99.8% and 99.9% identical to their human versions.

Additional to protein modelling, AI has also been used to identify the effect of existing drugs on COVID-19 through the prediction of protein-ligand binding affinity and molecular docking. Protein-ligand binding affinity is critical in tackling the drug repurposing problem. Drug repurposing can speed up the marketing of new drugs by combining virtual drug screening methods with ML approaches for the development of new drugs (Zhavoronkov et al. 2020). To identify novel drug candidates against COVID-19, Kadioglu et al. have combine virtual drug screening, molecular docking, and supervised ML techniques (Kadioglu et al. 2020). The authors have used a library of FDA-approved drugs to investigate their potential for repurposing as anti-SARS-CoV-2 drugs as well as two chemical libraries obtained from literature mining and the ZINC database. They selected three viral proteins, namely spike protein, nucleocapsid protein, and 2'-o-ribose-methyltransferase as targets for their combined approaches using blind docking mode in virtual screening, molecular docking, and supervised Naive Bayes and neural network. Their results have identified several FDA-approved drugs against hepatitis C, such as Paritaprevir, Semeprevir, Grazoprevir, and Velpatasvir, as candidates against COVID-19. Their results are also supported and validated by reports from Elfiky et al. (2017) that anti-HCV compounds are active against MERS-CoV. Stebbing et al. has also developed Benevolent AI's proprietary artificial intelligence (AI)-derived knowledge graph to identify drugs against SARS-CoV-2 (Stebbing et al. 2020). They identified a group of approved drugs that are members of the numb-associated kinase (NAK) family,including AAK1 and GAK; the inhibition of which has been shown to reduce viral infection of cells.

2.2.2 Fast Diagnosis Using Deep Learning Methods

One of the most popular areas where AI being applied to COVID-19 response has focused on fast detection based on medical data collector from radio imaging and biosensor (Ting et al. 2020). COVID-19 exhibits unique radiological signatures and image patterns that can be observed in medical imagery (Ng et al. 2020). Ai et al. (2020) have identified that 59% of the 1014 COVID-19 positive patient data in China had shown both positive RT-PCR results and chest CT scans and 88% (888/1014) for the diagnosis of suspected patients with COVID-19, respectively. The sensitivity of chest CT imaging for COVID-19 was 97% and can be considered as a primary tool for the current COVID-19 detection in epidemic areas. However, due to the large number of lung CT and X-Ray scans, it can be time-consuming for radiologists to identify these patterns. As a result, ML approaches have been adopted in the CT scan for the detection (Pirouz et al. 2020).

Bai et al. (2020) have proposed COVID-19 detection neural network (COVNet) to detect COVID-19, using a three-dimensional deep learning neural networks on three types of chest CT imaging data, namely COVID-19, Community Acquired Pneumonia (CAP), and other non-pneumonia. Max pooling operation is applied to combine features extracted from the CT scan by ResNet-50 CNNs, before these features are fed into a fully connected CNN to compute probabilities of the three data types. Their results have shown COVNet can achieve a high accuracy in detecting COVID-19 cases. Other approaches, such as Inception and UNet, have also been applied. Gozes et al. (2020) have implemented a two-step approach to train the medical images, using robust 2D and 3D deep learning model. They have used the UNet architecture to identify the area of interest on the lung and ResNet-50 to train and fine-tuned the model to classify the images as positive or negative. Their systems have accelerated the detection and diagnostic of new cases.

Other than using medical images for detections, Imran et al. (2020) have developed AI4COVID-19, an AI-based test for remote preliminary diagnosis based on 2 second cough sampling. The app analyzed the cough samples, using an AI engine running on the cloud. They have shown that the AI4COVID-19 can distinguish between COVID-19 patient and others non-COVID-19 cough using novel multi-pronged mediator-centered risk-averse AI architecture to minimize misdiagnosis.

2.2.3 Pandemic Tracking Using Sensor Data Analytic

Artificial intelligence has played a major role in the development of application for pandemic tracking with the vast number of populations. The ability to provide fast and accurate information on the pandemic has played important roles in controlling and predicting the spread of COVID-19. This information can help to predict the risked people being infected or the outcome from the infection, mitigating the burden on the healthcare systems, as the demands for treatment increased. Artificial intelligence has been used to produce epidemiological models to predict the total number of confirmed cases, transmission rate, test accuracy, and death rate (Saurabh and Sougat 2020). These algorithms have been applied on patient information and daily activities to learn and predict the infection risk of individual. Many countries, including Brunei Darussalam, have adopted the AI systems for virus containment through social app. These models have helped many countries to manage the spread and flattening the curve.

Dong et al. from Johns Hopkins University, Baltimore, USA have developed a real-time interactive web-based dashboard to visualize and track the reported confirmed cases, including the city location and countries affected (Dong et al. 2020). The authors highlight that data sharing is vital to evaluate and maintain accurate reporting and reduce the amount of fake news. A hybrid AI model is used to analyze the level of infections of a virus carrier within the first few days after infection (Du et al. 2020). The model combines the Susceptible Infection Model, NLP, and the long short-term memory (LSTM) to forecast the COVID-19 infection rate. This hybrid model has identified that new confirmed cases are mainly infected by confirmed cases within 3–8 days and the average infection time is about 5.5 days.

Due to large volume of social data available, other ML algorithms have been used to classify the data and make them more useful. Hybrid neural network has been used to categorize spatiotemporally explicit cellphone mobility data as surrogate markers for physical distancing, along with population-weighted density and other local data points using epidemiological model (Soures et al. 2020). Lopez et al. (2020) applied NLP and text mining to analyze a multilanguage Twitter dataset to understand changing policies and common responses to the COVID-19. The systems can help to mitigate the impacts of the current pandemic and allow the government to prepare better actions for possible future pandemics. Support vector machine, Naive Bayes, and random forest are used to classify 3,000 posts collected from Sina Weibo into 7 types of situational information that can help to identify and predict the propagation scale, sense the mood of the public, and understand the situation during the crisis (Li et al. 2020). Allam and Jones recommended that a standardized protocol for the use of AI and data sharing is required to provide a better understanding and management of urban health during any pandemic, including the COVID-19 (Allam and Jones 2020).

3   Conclusion and Future Outlook

Biosensors are useful point-of-care diagnostic tools that could be used not only for surveillance purpose, but also to study the characteristics and behaviors of the novel COVID-19 virus. The applications of biosensor in real clinical settings are gradually exiting the proof-of-concept, and with the integration of nanotechnology and supporting technology, this will revolutionize diagnostic technology especially during a crisis. With AI, its real-world usage has been focusing on remote virus tracking and containment. Among the published works, the use of AI deep learning techniques for COVID-19 detection and viral progression based on radiology imaging and biosensor data appears to be dominant. Artificial intelligence has also been used to manage social distancing policies, lockdown, medical consultation, and other control measure. In computational biology, AI has helped the scientist to understand COVID-19 or discover novel drug compounds against the virus through the investigation of genetics and structure of the virus. However, there are still plenty of works need to be done for the use of AI in biological computation models due to the limited amount of data available in that area about COVID-19. It is hope that with the availability of data obtained from biosensors, these AI algorithms can be benchmarked using a structural framework to validate and compare the results.

References

Abbott laboratories (2020) Website at https://www.abbott.com/coronavirus.html

Ahmed SR, Kang SW, Oh S, Lee J, Neethirajan S (2018) Chiral zirconium quantum dots: a new class of nanocrystals for optical detection of coronavirus. Heliyon, e007

Ai T, Yang Z, Hou H, Zhan C, Chen C, Lv W, Tao Q, Sun Z, Xia L (2020) Correlation of chest CT and RT-PCR testing in coronavirus disease 2019 (COVID-19) in China: a report of 1014 cases. Radiology 296:E32–E40

Allam Z, Jones DS (2020) On the coronavirus (COVID-19) outbreak and the smart city network: universal data sharing standards coupled with artificial intelligence (AI) to benefit urban health monitoring and management. Healthcare 8(1):46

Bai X, Hsieh B, Xiong Z, Halsey K, Choi JW, Tran T, Pan I, Shi L, Wang D, Mei J, Jiang X, Zeng Q, Egglin T, Hu P, Agarwal S, Xie F, Li S, Healey T, Atalay M, Liao W (2020) Performance of radiologists in differentiating COVID-19 from viral pneumonia on chest CT. Radiology 296:E46–E54

Bezinge L, Suea-Ngam A, deMello AJ, Shih CJ (2020) Nanomaterials for molecular signal amplification in electrochemical nucleic acid biosensing: recent advances and future prospects for point-of-care diagnostics. Mol Syst Des Eng 5:49–66

Cascella M, Rajnik M, Cuomo A, Dulebohn SC, Di Napoli R (2020) Features, evaluation and treatment coronavirus (COVID-19). In StatPearls [Internet]. StatPearls Publishing

Chen S-H, Chuang Y-C, Lu YC, Lin H-C, Yang YL, Lin CS (2009) A method of layer-by-layer gold nanoparticle hybridization in a quartz crystal microbalance DNA sensing system used to detect dengue virus. Nanotechnology 20:215501

Chen S, Yang J, Yang W, Wang C, Bärnighausen T (2020) COVID-19 control in China during mass population movements at new year. Lancet 395(10226):764–766

Cooper MA (ed) (2009) Label-free biosensors: techniques and applications. Cambridge University Press, Cambridge/New York

Dong E, Du H, Gardner L (2020) An interactive web-based dashboard to track COVID-19 in real time. Lancet 20:533–534

Draz MS, Shafiee H (2018) Applications of gold nanoparticles in virus detection. Theranostics 8:1985–2017. https://doi.org/10.7150/thno.23856

Du S, Wang J, Zhang H, Cui W, Kang Z, Yang T, Lou B, Chi Y, Long H, Ma M, Yuan Q, Zhang S, Zhang D, Xin J, Zheng N (2020) Predicting COVID-19 Using Hybrid AI Model. Available at SSRN: https://ssrn.com/abstract=3555202

Elfiky AA, Mahdy SM, Elshemey WM (2017) Quantitative structure-activity relationship and molecular docking revealed a potency of anti-hepatitis C virus drugs against human corona viruses. J Med Virol 89:1040–1047

Esteban C, Hyland S, Ratsch G (2017) Time series generation with recurrent conditional GANs. arXiv preprint arXiv:1706.02633

Fahmiin MA, Lim TH (2020) Evaluating the effectiveness of wrapper feature selection methods with artificial neural network classifier for diabetes prediction. Lecture notes of the institute for computer sciences, social informatics and telecommunications engineering, vol 309. Springer, Cham.

Generalov VM, Naumova OV, Fomin BI, P'Yankov SA, Khlistun IV, Safatov AS, Aseev AL (2019) Detection of Ebola virus VP40 protein using a nanowire SOI biosensor. Optoelectron Instrum Data Process 55:618–622. https://doi.org/10.3103/s875669901906013x

Gozes O, Frid-Adar M, Greenspan H, Browning PD, Zhang H, Ji W, Bernheim A, Siegel E (2020) Rapid AI development cycle for the coronavirus (COVID-19) pandemic: initial results for automated detection & patient monitoring using deep learning CT image analysis. arXiv:2003.05037

Heymann DL, Shindo N (2020) COVID-19: what is next for public health? Lancet 395 (10224):542–545

Huang JC, Chang YF, Chen KH, Su LC, Lee CW, Chen CC, ... Chou C (2009) Detection of severe acute respiratory syndrome (SARS) coronavirus nucleocapsid protein in human serum using a localized surface plasmon coupled fluorescence fiber-optic biosensor. Biosens Bioelectron 25:320–325

Imran A, Posokhova I, Qureshi HN, Masood U, Riaz S, Ali K, John C, Nabeel M (2020) AI4COVID-19: AI enabled preliminary diagnosis for COVID-19 from cough samples via an app. arXiv:2004.01275

Ishikawa FN, Chang HK, Curreli M, Liao HI, Olson CA, Chen PC, ... Zhou C (2009) Label-free, electrical detection of the SARS virus N-protein with nanowire Biosensors utilizing antibody mimics as capture probes. ACS Nano 3:1219–1224. https://doi.org/10.1021/nn900086c

Jumper J, Tunyasuvunakool K, Kohli P, Hassabis D, AlphaFold Team (2020) Computational predictions of protein structures associated with COVID-19. DeepMind website. 2: https://deepmind.com/research/open-source/computational-predictions-of-protein-structures-associated-with-COVID-19

Kadioglu O, Saeed M, Johannes G, Efferth T (2020) Identification of novel compounds against three targets of SARS CoV-2 coronavirus by combined virtual screening and supervised machine learning. Bull World Health Organ. E-pub

Kaya SI, Karadurmus L, Ozcelikay G, Bakirhan NK, Ozkan SA (2020) Electrochemical virus detections with nanobiosensors. In: Nanosensors for smart cities. Elsevier, San Diego, pp 303–326

Kim H, Park M, Hwang J, Kim JH, Chung DR, Lee KS, Kang M (2019) Development of label-free colorimetric assay for MERS-CoV using gold nanoparticles. ACS Sens 4:1306–1312. https://doi.org/10.1021/acssensors.9b00175

Li L, Zhang Q, Wang X, Zhang J, Wang T, Gao TL, Duan W, Tsoi KK, Wang FY (2020) Characterizing the propagation of situational information in social media during COVID-19 epidemic: a case study on Weibo. IEEE Trans Comput Soc Syst 7:56–562

Lim SA, Ahmed MU (2016) Electrochemical immunosensors and their recent nanomaterial-based signal amplification strategies: a review. RSC Adv 6:24995–25014

Lim SA, Ahmed MU (2019) Chapter 1: Introduction to Immunosensors. In: Immunosensors. Royal Society of Chemistry, London, pp 1–20

Lopez CE, Vasu M, Gallemore C (2020) Understanding the perception of COVID-19 policies by mining a multilanguage Twitter dataset. arXiv preprint arXiv:2003.10359

Mahapatra S, Chandra P (2020) Clinically practiced and commercially viable nanobio engineered analytical methods for COVID-19 diagnosis. Biosens Bioelectron 2020(165):112361

Mao K, Zhang H, Yang Z (2020) Can a paper-based device trace COVID-19 sources with wastewater-based epidemiology? Environ Sci Technol 2020(54):3733–3735

Monosik R, Streansky M, Surdik E (2012) Biosensors-classification, characterization and new trends. Acta Chim Slov 5:109–120

Ng M, Lee YP, Yang J, Yang F, Li X, Wang H, Lui M, Lo S, Leung B, Khong PL, Hui KM, Yuen K, Kuo MD (2020) Imaging profile of the COVID-19 infection: radiologic findings and literature review. Radiol Cardiothorac Imaging 2:1

Pirouz B, Shaffiee Haghshenas S, Piro P (2020) Investigating a serious challenge in the sustainable development process: analysis of confirmed cases of COVID-19 (new type of coronavirus) through a binary classification using artificial intelligence and regression analysis. Sustainability 12:2427

Qiu G, Gai Z, Tao Y, Schmitt J, Kullak-Ublick GA, Wang J (2020) Dual-functional plasmonic photothermal biosensors for highly accurate severe acute respiratory syndrome coronavirus 2 detection. ACS nano

Reza SM, Najarian K (2016) Transforming big data into computational models for personalized medicine and health care. Dialogues Clin Neurosci 18:339–343

Rinken T, Kivirand K (2019) Biosensors for environmental monitoring

Rothan HA, Byrareddy SN (2020) The epidemiology and pathogenesis of coronavirus disease (COVID-19) outbreak. J Autoimmun 109:102433

Saurabh B, Sougat R (2020) Going viral – Covid-19 impact assessment: a perspective beyond clinical practice. J Mar Med Soc 22(1):9–12

Saylan Y, Erdem Ö, Ünal S, Denizli A (2019) An alternative medical diagnosis method: Biosensors for virus detection. Biosensors 9:65

Senior A, Evans R, Jumper J et al (2019) Protein structure prediction using multiple deep neural networks in the 13th critical assessment of protein structure prediction (CASP13). Proteins 87:1141–1148

Senior A, Evans R, Jumper J, Kirkpatrick J, Sifre L, Green T, Qin C, Žídek A, Nelson A, Bridgland A, Penedones H, Petersen S, Simonyan K, Crossan S, Kohli P, Jones D, Silver D, Kavukcuoglu K, Hassabis D (2020) Improved protein structure prediction using potentials from deep learning. Nature 7792:706–710

Seo G, Lee G, Kim MJ, Baek SH, Choi M, Ku KB, Kim SJ (2020) Rapid detection of COVID-19 causative virus (SARS-CoV-2) in human nasopharyngeal swab specimens using field-effect transistor-based biosensor. ACS nano

Shi L, Sun Q, He JA, Xu H, Liu C, Zhao C, . . . Long J (2015) Development of SPR biosensor for simultaneous detection of multiplex respiratory viruses. Biomed Mater Eng 26(s1):S2207–S2216

Shi F, Wang J, Shi J, Wu Z, Wang Q, Tang Z, He K, Shi Y, Shen D (2020) Review of artificial intelligence techniques in imaging data acquisition, segmentation and diagnosis for COVID-19. IEEE Rev Biomed Eng 1

Soures N, Chambers D, Carmichael Z, Daram A, Shah DP, Clark K, Potte L, Kudithipudi D (2020) SIRNET: understanding social distancing measures with hybrid neural network model for COVID-19 infectious spread, arXiv:2004.10376

Stebbing J, Phelan A, Griffin I, Tucker C, Oechsle O, Smith D, Richardson P (2020) COVID-19: combining antiviral and anti-inflammatory treatments. Lancet Infect Dis 20-4.400–402

Steinmetz M, Lima D, Viana AG, Fujiwara ST, Pessôa CA, Etto RM, Wohnrath K (2019) A sensitive label-free impedimetric DNA biosensor based on silsesquioxane-functionalized gold nanoparticles for Zika virus detection. Biosens Bioelectron 141:111351. https://doi.org/10.1016/j.bios.2019.111351

Stroock AD (2008) Microfluidics. In: Optical biosensors. Elsevier, pp 659–681. Ligler FS, Taitt CR (eds) (2011). Optical biosensors: today and tomorrow. Elsevier

Takhistov P (2005) Biosensor technology for food processing. Saf Packag. https://doi.org/10.1201/b15995-143

Tam PD, Hieu NV, Chien ND, Le A-T, Tuan MA (2009) DNA sensor development based on multi-wall carbon nanotubes for label-free influenza virus (type a) detection. J Immunol Methods 350 (1–2):118–124. https://doi.org/10.1016/j.jim.2009.08.002

Teengam P, Siangproh W, Tuantranont A, Vilaivan T, Chailapakul O, Henry CS (2017) Multiplex paper-based colorimetric DNA sensor using pyrrolidinyl peptide nucleic acid-induced AgNPs aggregation for detecting MERS-CoV, MTB, and HPV oligonucleotides. Anal Chem 89 (10):5428–5435

Ting DSW, Carin L, Dzau V, Wong TY (2020) Digital technology and COVID-19. Nat Med 26:459–461

Tuena C, Semonella M, Fernández-Álvarez J, Colombo D, Cipresso P (2019) Predictive precision medicine: towards the computational challenge. In: Pravettoni G, Triberti S (eds) P5 eHealth: an agenda for the health technologies of the future. Springer

Wei G (2019) Protein structure prediction beyond AlphaFold. Nat Mach Intell 1:336–337

Weng X, Neethirajan S (2018) Immunosensor based on antibody-functionalized MoS 2 for rapid detection of avian coronavirus on cotton thread. IEEE Sensors J 18:4358–4363

World Health Organization (2020) Accessed 17 Apr 2020 at 8.28 pm, https://www.who.int/emergencies/diseases/novel-coronavirus-2019

Wu A, Khan WS (eds) (2020) Nanobiosensors: from design to applications. Wiley, Weinheim

Wynants L, Van Calster B, Bonten M, Collins GS, Debray T, De Vos M, Haller M, Heinze G, Moons K, Riley R, Schuit E, Smits L, Snell K, Steyerberg W, Wallisch C, Smeden M (2020)

Prediction models for diagnosis and prognosis of covid-19 infection: systematic review and critical appraisal. Br Med J 369:m1328

Zhang C, Zheng W, Huang X, Bell E, Zhou X, Zhang Y (2020) Protein structure and sequence reanalysis of 2019-nCoV genome refutes snakes as its intermediate host and the unique similarity between its spike protein insertions and HIV-1. J Proteome Res 19:1351–1360

Zhao CT, Daubinger P (2019) Nanobiosensors. Johnson Matthey Technol Rev 63:205–207

Zhavoronkov A, Aladinskiy V, Zhebrak A, Zagribelnyy B, Terentiev V, Bezrukov D, Polykovskiy D, Shayakhmetov R, Filimonov A, Orekhov P, Yan Y, Popova O, Vanhaelen Q, Aliper A, Ivanenkov Y (2020) Potential COVID-2019 3C-like protease inhibitors designed using generative deep learning approaches. Chemrvix website, https://chemrxiv.org/articles/Potential_2019-nCoV_3C-like_Protease_Inhibitors_Designed_Using_Generative_Deep_Learning_Approaches/11829102

Diverse Molecular Techniques for Early Diagnosis of COVID-19 and Other Coronaviruses

Sharmili Roy and Anupriya Baranwal

Abstract

Novel coronavirus (SARS-CoV-2) pandemic has impacted the world severely causing millions of positive cases and thousands of fatality cases. Admission of a large number of infected patients in the hospitals is creating daily challenges for medical experts and researchers, such as limited numbers of isolation rooms, less amount of personal protective equipment (PPE), maintaining the hygiene in the hospitals, handling of many clinical samples for tests at the laborites, etc. This disease is mild in most people, but in pre-medical conditions and some elderly people, this virus is affecting majorly and causing fatality. Symptoms of this viral infection include pneumonia, acute respiratory diseases, and multi-organ dysfunction. Considering the situation, early diagnosis of the disease is extremely crucial to control the ongoing epidemic. In this chapter, we have summarized current molecular assays that are in use at many laboratories and hospitals for coronavirus detections. We also discussed molecular diagnosis cases of Severe Acute Respiratory Syndrome Coronavirus (SARS-CoV) and the Middle East Respiratory Syndrome Coronavirus (MERS-CoV) viruses. Many improved and new approaches of molecular tools are involved for the detection of these viruses that are listed down in the table form. We have also reviewed the current challenges/issues arising currently for the effective diagnostics and treatment of the pandemic situation. Further, we put heads together about the future directions

S. Roy (✉)
Division of Oncology, School of Medicine, Stanford University, Stanford, CA, USA

A. Baranwal
Sir Ian Potter NanoBiosensing Facility, NanoBiotechnology Research Laboratory, School of Science, RMIT University, Melbourne, VIC, Australia
e-mail: sharmili@stanford.edu

P. Chandra, S. Roy (eds.), *Diagnostic Strategies for COVID-19 and other Coronaviruses*, Medical Virology: From Pathogenesis to Disease Control,
https://doi.org/10.1007/978-981-15-6006-4_7

and necessitating investigations that are needed for developing more effective detection of coronavirus infections.

Keywords

Coronavirus · COVID-19 · Molecular diagnosis · RT-PCR · RT-LAMP-based methods · Pandemic · Sample types

1   Introduction

In December 2019, a cluster of patients was hospitalized due to unknown causes in Wuhan, China with an initial diagnosis of viral pneumonia (Velavan and Meyer 2020). Soon, this disease became an outbreak and started spreading rapidly into different parts of the world with respiratory droplets being a common medium for a human to human transmission. Most of the cases, the elderly persons started getting affected severely. Though the disease symptoms are mild in young and healthy peoples, it affects majorly the old and immune-compromised persons. Due to the rapid spreading and increasing number of fatal cases around the world, the World Health Organization (WHO) announced this outbreak as a pandemic and a global health emergency (WHO, https://www.who.int/). Initial studies pointed the origin of this viral disease to a seafood market in Wuhan, with the possible transmission of the infectious virus from an animal to humans, causing a massive outbreak later (Riou and Althaus 2020; Zhou et al. 2020a). The virus named "2019-nCoV" by the WHO, which was later renamed by the International Committee on Taxonomy of Viruses as "Severe Acute Respiratory Syndrome Coronavirus 2" (SARS-CoV-2) or COVID-19. The SARS-CoV-2 is a single-stranded RNA (+ssRNA) beta subtype of coronavirus (Latin: Corona – crown) surrounded by spike protein (glycoproteins in nature) that mediates the virus entry into host cells in the human body, causing common cold to severe acute respiratory disease (Ibrahim et al. 2020; Weiss and Navas-Martin 2005; Zhou et al. 2020b). To date (05/25/2020), the total death due to COVID-19 is 347,563 worldwide with an unfortunate possibility of increasing further without proven therapeutics and a vaccine (worldometers.info/coronavirus/USA). Therefore, early detection of COVID-19 infection has become very critical to check the spread of the disease. A rapid testing kit cannot only detect the infected one but also can help to monitor the stage and help in further medication. Currently, the nucleic acid-based tests and genetic sequencing are the ones used to detect COVID-19 infections (Li et al. 2020a), such as reverse transcriptase polymerase chain reactions (RT-PCR) (Chan et al. 2020), reverse transcriptase loop-mediated isothermal amplification (RT-LAMP) (Park et al. 2020), rolling circle amplification (RCA), microarray, next-generation sequencing (NGS) technologies (Gu et al. 2020; Pan et al. 2020), etc. (Fig. 1).

RT-PCR is considered to be a gold standard diagnostic approach due to its accuracy, high specificity, and sensitivity (Tahamtan and Ardebili 2020). RT-LAMP is used for its straight forward and simple methodology. In this chapter, we have discussed in detail about the sample collection and processing guidelines,

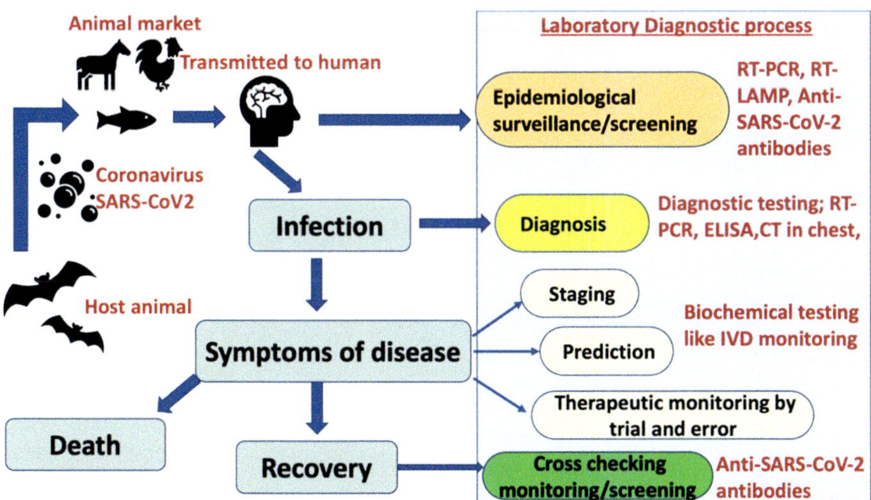

Fig. 1 The flowchart of COVID-19 that explains how the virus is originated from animals, circulated in the wet seafood market, and lastly transmitted to the human population. After transmission into humans, it incubates inside the body for 2–14 days. Meanwhile, epidemiological screening was performed at the laboratory, which includes RT-PCR, RT-LAMP, ELISA, etc. Ensuring the positive results from patients, they are kept in isolation, and various diagnostic processes were performed for therapeutic monitoring. Once all symptoms are reduced, then they were tested again to confirm the trace of the virus inside the body

the nucleic acid-based diagnostic methodologies, their pros and cons, and finally the future directions for diagnostic research.

2 Sample Processing

The human-to-human COVID-19 transmission is very fast with the possibility of a single person infecting three other people via respiratory droplets as well as another medium (Liu et al. 2020). The virus has an incubation period of 2–14 days, post which the symptoms can vary from person to person (Li et al. 2020c; Udugama et al. 2020). The symptoms of the disease resemble that of common cold and influenza but with more deadly consequences making the safe collection and processing of diagnostic samples very critical (Fig. 2). Firstly, samples were collected from infected people in the form of throat swabs, nose swabs, serum, stools, etc. These samples were collected with a synthetic tip and an aluminum/plastic shaft. Samples should be kept in a proper condition in a tightly sealed container with 2–3 ml of viral transport media, so that it could not spread while transporting (Giuseppe et al. 2020). Later samples were transferred to research or medical laboratories for further analysis by molecular technique and sequencing for further research purposes. It is essential to keep the infected person completely isolated and under observation at a hygienic place following guidelines recommended by the WHO and centers for disease control and prevention (CDC).

Patient Workflow

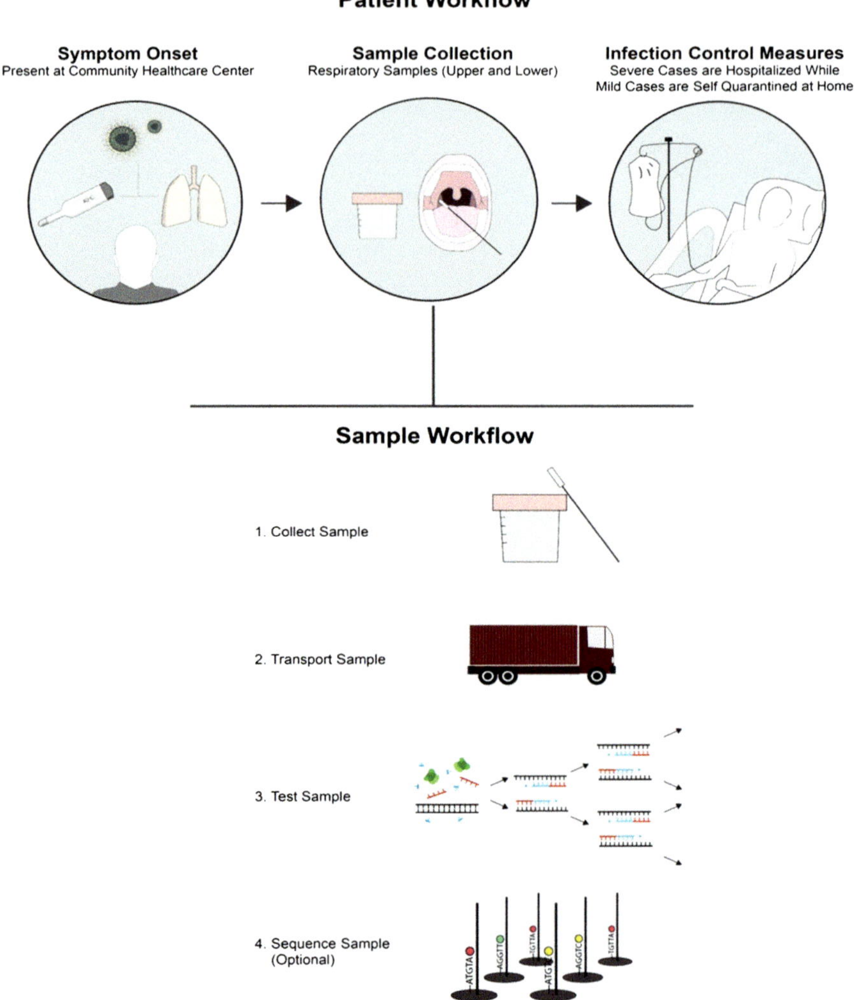

Fig. 2 Schematic workflow of sample processing during the COVID-19 epidemic. First, clinical samples were collected as throat swabs, serum, stools, etc. Then transfer those samples to prospective laboratories for further analysis by molecular testing and sequencing process. (This figure is reproduced with permission from Udugama et al. (2020))

3   Molecular Diagnostic Techniques

The molecular diagnostic is a collection of different nucleic acid-based methodologies that analyzes numerous biological markers. Molecular assays are advanced tools that made diagnostic path easy and straightforward. The current COVID-19 pandemic desires early and onsite detection technology, and these molecular tools can contribute majorly for early detection. Concisely, we have hoisted up a few case studies as an example where these molecular tools have been put in use to diagnose COVID-19. Previously, these processes were also reported to be used to diagnose SARS and MERS during their outbreaks.

3.1 Nucleic Acid Detection Technology

Nucleic acid amplification (NAA) testing process is one of the main and commonly used molecular tools for current COVID-19 screening. As mentioned above, five different technologies are in use for all kinds of coronavirus detection processes. These tests reveal the trace of viruses qualitatively and quantitatively. Depending on these NAA technologies, researchers are now able to determine viral presence from the collected clinical samples, like saliva, blood, sputum, or any respiratory swabs. Simultaneously, these technologies are very specific due to the usage of specifically designed primers and effortless compared with conventional processes (Mothershed and Whitney 2006; Bloomfield et al. 2015). Apart from these advantages, these techniques are also helpful for real-time monitoring of the disease as well (Klein 2002; Shen et al. 2020).

3.1.1 RT-PCR-Based Methodology

Polymerase chain reaction (PCR) is a molecular technique that is used to make millions and billions copies of specific DNA specimens with the help of specific primers. This method has the capacity of determining many parameters, like higher sensitivity, higher accuracy, higher specificity, and higher reproducibility of coronaviruses (Balboni et al. 2012). Coronaviruses have ssRNA as their genetic material, therefore to perform PCR, first, the RNA should be converted into a complementary DNA (cDNA) following reverse transcription protocol. Later, PCR amplification is performed on these cDNA templates by using specifically designed primers. The amplicons from cDNA samples run on a gel followed by its visualization of bands in Gel Doc or E-Gel imager (Setianingsih et al. 2019). However, the advanced quantitative real-time PCR (qRT-PCR) does not require any post-PCR processing, like gel electrophoresis for visualization of the positive bands. All results are then obtained from the software with proper sensitivity values. Currently, this process is very much favorable for less laborious, high-throughput screening in hospitals and research laboratories for COVID-19 diagnostics (Corman et al. 2012; Chan et al. 2020). The main aim of RT-PCR is to detect the presence of RNA specific to the SARS-CoV-2 virus which is responsible for the occurrence of COVID-19 disease. Clinical samples are processed for RNA isolation and purification by using commercially available RNA isolation kits. After RNA isolation, specific primers, fluorescent probes, and DNA polymerase were added for further PCR amplification reactions. In real-time RT-PCR process, two types of dyes are needed; one is a quencher and another one as a reporter (Nagaraj et al. 2006). Fluorescent probes emit a signal that increases when more copies of DNA are created. If the test is positive, the fluorescence level crosses a definite threshold

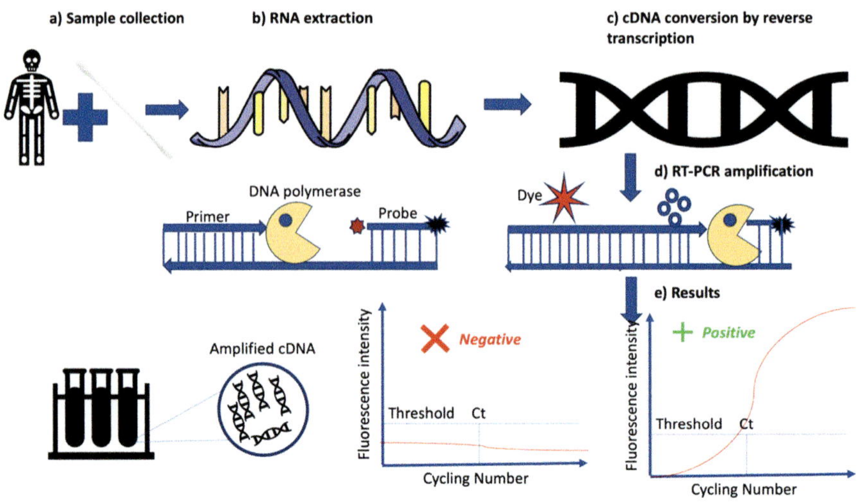

Fig. 3 Real-time RT-PCR steps while testing clinical samples: (**a**) sample collection from throat or nose of infected person (**b**) RNA isolation from processed and purified clinical samples (**c**) cDNA production by reverse transcriptase (**d**) RT-PCR amplification by thermal cycler (**e**) test result analysis of cDNA amplicons from RT-PCR amplification process

point beyond the estimated background levels. If the test is negative, i.e., the virus is not present in the test sample, then the PCR will not amplify cDNA. As a result, the fluorescence level will not reach the estimated threshold point (Fig. 3) (Ni et al. 2012). The most important feature of RT-PCR assay is to allow real-time quantification of DNA copy numbers of the targeted virus with high sensitivity and specificity.

Further development for the detection of COVID-19 by RT-PCR improvised with synthetic nucleic acid technology for robust diagnosis. At first, RNA was extracted from the clinical samples using viral RNA extraction kit, and RT-PCR reaction in thermal cycler was performed. In this study, the sensitivity of the SARS-CoV-2 genomic DNA was determined to be 3.9 copies per reaction for the E-Gene and 3.6 copies per reaction for the RdRp gene, respectively. Further, the specificity was compared with other coronaviruses and bacterial strains in clinical samples and with negative human sputum samples. None of the other strains from different viruses showed cross-reactivity, following this assay (Corman et al. 2020). On the other hand, a group of scientists from Wuhan, China, reported about constant monitoring by RT-PCR of positive test results from patients who recovered from COVID-19. These patients were all medical professionals and aged between 30–36 years. The type of symptoms was matched with a mild acute respiratory problem, such as fever, cough, and shortness in breathing problems. They were treated with the proper quarantine protocol for about 12 days, and their test results were evaluated by using RT-PCR. The patients recovered after 32 days; however, they were called back for re-test of COVID-19 before joining back their usual schedules. PCR can confirm COVID-19 cases repeatedly and help to monitor them from time to time for complete confirmation (Lan et al. 2020). In this chapter, we have summarized a few more cases monitored by RT-PCR assays for various coronaviruses in detail in Table 1. Like any other assay, RT-PCR is not devoid of demerits and suffers from issues of contamination, high cost, time-consuming analysis, exhaustive sample preparation, the requirement for trained personnel and high-end instrumentation, etc. (van Elden et al. 2004). To overcome the demerits of RT-PCR, another nucleic acid-based alternative diagnostic system LAMP is used.

3.1.2 LAMP-Based Methodology

Loop-mediated isothermal amplification (LAMP) is a novel, rapid, and low cost molecular diagnostic approach that offers high accuracy and high sensitivity. The working principle of the LAMP process is based on auto-cycling strand displacement DNA amplification by *Bst* DNA polymerase. This process is called isothermal, because the whole molecular reaction occurs at one single optimized temperature between 60–65 °C with six specific primers; it does not need several cycles and temperature, likse conventional PCR process (Notomi et al. 2000; Roy et al. 2016a, c; Roy et al. 2017a; Azam et al. 2018). This process requires only a small heat block or water bath with temperature regulation which makes this process portable and perfect for onsite analysis. Besides, this technique does not require much time to finish the whole process, it takes around 1 h or less to obtain any kind of qualitative and quantitative data (Roy et al. 2016b; Roy et al. 2017b). Based on these emerging characteristics of the LAMP, researchers are more relying to this process for the detection of COVID-19 for rapid alternative outcomes (Enosawa et al. 2003). The process is divided into two phases where two types of elongation reactions occur, a cyclic reaction and a non-cyclic amplification reaction. The first phase is a cyclic step. Earlier in this stage, the template DNA strand is hybridized with forwarding inner primers (FIP) at 3′ terminal; this process is called self-elongation from the stem-loop structure. During this phase, F3 primer anneals to F3c region in the sequence. Due to this, it starts synthesizing the complementary DNA strand. Each of the inner primers acquires a sequence that is complementary to one chain for amplification at the 3′ terminal and matching to the inner region of the similar chain at the 5′ terminal. Later, the chain amplification starts with the help of *Bst* DNA polymerase enzyme that mediates strand displacement synthesis, and the reaction starts releasing FIP-linked complementary strand. It synthesizes double-stranded DNA from the F3 primer and the template DNA (Fig. 4). In the later stage, the complementary F1c and F1 region in the targeted sequence bind with the chain at 5′ end, and the stem-loop structure are created by the complementary base. Next, the same strand interacts with the backward inner primers (BIP) and B3 primers. The same reaction occurs at this stage, and another new loop structure creates at another end. Therefore, a dumbbell structure is produced, and this structure provides the template for reacted amplicons during the consequent reaction (Ushikubo 2004; Notomi et al. 2015; Li et al. 2017). In the second phase, as a non-cyclic amplification step, the formed dumbbell structure as the template and the dumbbell-like DNA structure convert into a loop DNA structure by a self-priming process of DNA

Table 1 Summary of detection of coronaviruses by using RT-PCR, RT-LAMP, and RCA assay as molecular diagnostic tools

Diagnostic tool	Sample type	Detection Time	LOD	Detection range	Original source for samples	Specificity testing	References
RT-PCR	Fecal sample from COVID-19 patient	Not mentioned	3.9 copies/ reaction for E-gene assay and 3.6 copies/ reaction for RdRp assay	Potential exposure to a common environmental source with increased sequence variability	Human and bat	Water as a negative sample; cross-reactivity test with other coronavirus and bacteria in clinical samples	Corman et al. (2020)
RT-PCR	Throat swabs, pleural fluid, blood samples from SARS outbreak in 2003	Not mentioned	3 genome copies/reaction	10^{-3} to 10^{-8} dilution [1–10 genome copies]	RNA from human coronaviruses	All type of coronaviruses strain; respiratory syncytial virus; parainfluenza, influenza A, adenoviruses and parainfluenza 3	Adachi et al. (2004)
Singleplex PCR and later sequenced the amplicons	Biological specimens, such as sputum, nasal swab and serum from MERS-CoV infections	Not mentioned	Not mentioned since it is a comparison study between different viral strains	Qualitative study	Human	Comparison study between different viral strain, including coronavirus, influenza virus, astrovirus, Nipah virus, adenovirus, hantavirus, etc.	Setianingsih et al. (2019)
RT-PCR	Snap-frozen autopsy samples from SARS patients	Not mentioned	5 copies/ reaction	Linear over a 6-log range of target concentrations	Human	100% specificity with other patients	Hadjinicolaou et al. (2011)
Real-time RT-PCR	Nasopharyngeal aspirate/swabs, throat swabs, sputum specimens from	Not mentioned	11.2 genomic RNA copies/ reaction of COVID-19	Upto ten-fold dilution	Human	Specificity check with SARS-COV2 with SARS-CoV and MERS-CoV	Chan et al. (2020)

Multiplex real-time RT-PCR	Throat swabs from coronavirus	Not mentioned	10 copies/μL	The detection range of this assay ten-fold serial dilutions containing 10^6-10^0	Human	Specificity check with influenza virus and other corona viral strains	Yu et al. (2008)
One step qRT-PCR	Stool, nasopharyngeal aspirate, and serum from SARS patients	Not mentioned	1 copy/reaction	Dynamic detection range 10^3-10^0	Human	Specificity check with influenza virus, norovirus, and other human coronaviruses	Jiang et al. (2004(and Yip et al. (2005)
Real-time RT-PCR	Serum, nasal swabs from MERS infection	Not mentioned	This study is qualitative (yes/no)	Qualitative study	Human, camels	Different viral samples from animals	Chu et al. (2014)
RT-PCR	Throat swabs from MERS infection	Not mentioned	Not mentioned	Detection range is 500 to 80,000 copies per swabs	Human	Specificity checked with non-MERS strains, MERS-CoV, and HCoVs strains	Drosten et al. (2014)
RT-LAMP	Respiratory samples obtained from COVID-19 infected people	15–40 min	10 copies/ μL	1,000,000 to 10 copies/μL	Human	Specificity checked with 7 similar coronaviruses, 2 influenza viruses, and 2 normal coronaviruses	Yu et al. (2020)
RT-LAMP	Serum, urine, saliva, nasal swab, and oral swab from COVID-19 patients	30 min	~1.02 fg	Ranging from 0.204 fg to 10 ng	Human	Checked specificity with a different strain of SARS, MERS, BtCoV/ MHV	Lamb et al. (2020)
LAMF	Nasopharyngeal aspirate samples from SARS patients	Not mentioned	10 copies/ reaction	Not mentioned	Human	Specificity checked with RSV, adenovirus, influenza viral strains, and HMPV	Poon et al. (2004)

(continued)

Table 1 (continued)

Diagnostic tool	Sample type	Detection Time	LOD	Detection range	Original source for samples	Specificity testing	References
RT-LAMP	Clinical samples from COVID-19 patients	30–50 mins	3 copies/reaction	10^6 to 1 copies/µL	Human	Specificity checked with human coronaviruses OC43, 229E, HKU-1, and NL63	Lu et al. (2020)
Colorimetric LAMP	Respiratory specimens, like swabs collected from COVID-19 patients	30 mins	4.8 copies/µL	A pilot study of ~120 million copies to ~120 copies at ten-fold intervals	Human	Not mentioned	Zhang et al. (2020)
LAMP	Sputum, serum, nose swab from infected coronavirus person	1 h	1 copy/reaction	Not mentioned	Human	Specificity checked with other coronavirus strains, like HCoC-229E, HCoV-HKU1, influenza, and adenovirus, echovirus 9, human rhinovirus	Pyrc et al. (2011)
RT-LAMP	Nasopharyngeal swab as clinical specimens from MERS patients	30 min	3.4 copies	$5*10^5$–$5*10^{-1}$ copies/reaction	Human	Specificity checked with other respiratory viral strains, HCoV strains, RSV viral stains, adenovirus, other SARS strains	Shirato et al. (2014)
RT-LAMP	SARS-CoV-2 RNA was isolated from infected MRC-5 cells	30 min	100 copies/reaction	5 to 1000 RNA copies/reaction	African green monkey and human	All SARS, MERS-CoV strain, and another human coronavirus	Jung et al. (2020) and Park et al. (2020)

Single step RT-LAMP	Nucleocapsid protein as clinical samples	60 min	500 copies/mL	$5*10^6$–$5*10^0$ cells/mL	Human	Other coronavirus strains, adenovirus, human influenza viral strains	Jiang et al. (2020)
RCA	Throat swabs, stool, and urine samples from infected patients by SARS-CoV	90 min	Below 100 copies/ reaction	Not mentioned	Human	Specificity checked with human genomic DNA as a nonspecific template	Wang et al. (2005a)
RCA	Biopsy samples, such as saliva, tear, blood, from infected patients from MERS virus	2 h	0.1×10^{-12} M	0.1×10^{-12} M to 10×10^{-12} M	Human	Selectivity test checked with non-target pathogen strands	Jung et al. (2016)

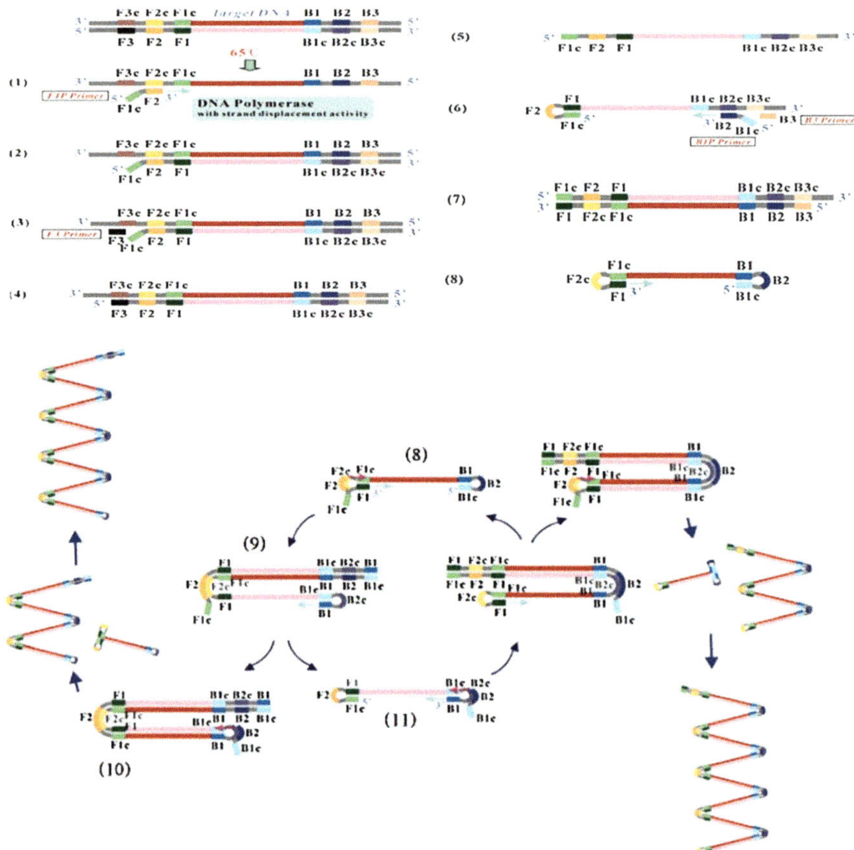

Fig. 4 Schematic representation of the LAMP reaction and its mechanism. Steps involved to prepare a dumbbell-shaped structure at both terminals. In this figure, both non-cyclic and cyclic amplification steps are elaborately illustrated with detailed mechanisms. (The figure is reproduced with permission from Ushikubo (2004) and Li et al. (2017))

synthesis. Following this, F2 and FIP hybridize the loop area of F2c of the amplicon. Concurrently, F1c collectively with F1 detaches the double-strand by releasing of one strand. At that time, the released single strand forms a loop structure at one end of the chain. Succeeding to this reaction, later loop structure forms at both the end by B1c and B1. Same way, B2 and BIP generate new DNA sequences by releasing one DNA strand (Lau et al. 2010). These repetitions continue until the final product construct with 10^9 copies as multiple loop DNAs.

LAMP has become an attractive diagnosis tool, mainly because of few reagents requirements, one constant temperature, high sensitivity, high accuracy, less involvement of instruments, less time consumption (~1 h), and less cost compared with PCR and other conventional procedures (Nagamine et al. 2001; Sheet et al. 2016).

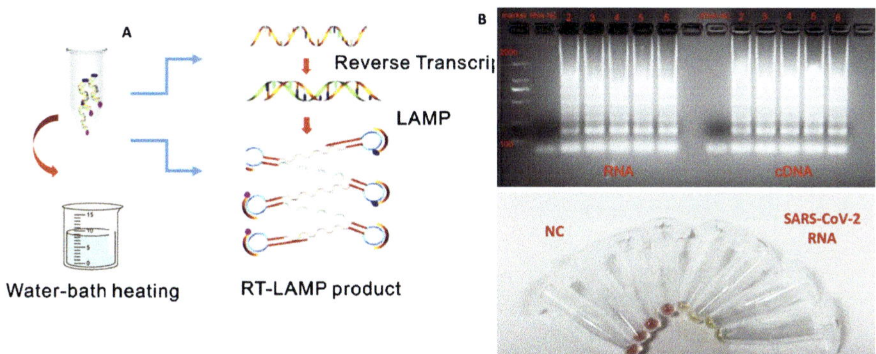

Fig. 5 Schematic representation of RT-LAMP with a colorimetric assay. (**a**) The strategy of amplification reaction for RT-LAMP (This figure is reproduced with permission from (Huang et al. 2018; Shen et al. 2020) (**b**) visualization of LAMP amplicon by agarose gel electrophoresis and color change in a single tube; positive samples that are containing SARS-CoV-2 RNA samples turned into pink to yellow to confirm the positive tests by adding phenol-red dye. (This figure is reproduced with permission from Yu et al. (2020))

A LAMP-based rapid detection test developed for COVID-19 testing, named as iLACO (isothermal LAMP based colorimetric method for CIVID 19), amplifies the ORF1ab gene using six specific primers (Yu et al. 2020). Twenty minutes of incubation at a normal water bath with newly designed RT-LAMP primers showed successful amplification which was confirmed by the change in sample color in the tubes. Hence, by the means of colorimetric reaction, the positive reaction can be confirmed without any involvement of high-end instruments. The color changed from pink to light yellow for a positive reaction in the tube, and meanwhile negative reactions did not change the color and remained as pink color. For the color change, they have used phenol-red dye (Fig. 5). The limit of detection (LOD) was found to be as low as around 10 copies/µL. Since qRT-PCR takes 2–3 h and need a more controlled environment with the high-end experimental procedure, iLACO process is much preferable in terms of rapidity and high sensitivity and accuracy. Another group diagnose COVID-19 using the LAMP method, just within 30 min with a low detection limit of ~1.02 fg and decent accuracy. For cross contamination experiment, they have tested same COVID-19 samples with other viral strains from SARS, MERS, and BtCoV, and it showed positive for COVID-19 containing samples and negative results for others, which evidenced the high specificity of this process (Lamb et al. 2020).

Previously, Shirato et al. reported to diagnose MERS-CoV using RT-LAMP assay. They have recorded the detection limit of 3.4 copies/reaction with no cross-reaction with other viral strains (Shirato et al. 2014). On the other hand, in the year 2011, Pyrc et al. reported the detection of human coronavirus strain-NL63 (HCoV-NL63) with LAMP assay. They have determined a very promising detection limit which was 1 copy/ reaction of RNA template. To confirm the successful assay performance, they run agarose gel electrophoresis with other strains, where they have found the specificity of the system. Researchers also reported that they have performed LAMP assay using a water bath (Pyrc et al. 2011). Referring to the accounted examples (Table 1), it can be concluded that LAMP assay is much more sensitive, cost-effective, time-efficient, highly accurate, and very specific diagnosis tool which promises a high-throughput screening for COVID-19.

3.1.3 RCA-Based Methodology

Rolling circle amplification (RCA) is also another kind of isothermal nucleic acid and signal-based amplification technique for the rapid detection of specific nucleic acid sequences. This technique has also appealed considerable attention in various disease diagnosis processes. First, RCA was reported by Nilsson et al., to detect target DNA (Nilsson et al. 1994), which provides a promising LOD with good specificity and accuracy (Chapin and Doyle 2011; Xu et al. 2019).

Rolling circle amplification is also capable of detection of coronaviruses by 10^9 fold within 90 min (Wang et al. 2005b). The main advantage of RCA-based detection involves isothermal condition with minimal reagents requirement and high level of sensitivity with less false-positive outcomes which is highly comparable with conventional processes, like PCR. This diagnosis process involves two priming approaches. The first approach uses a single primer, and the other uses pair of primers. Single-primed RCA results in a linear chain reaction in the monotonous sequence of DNA. Generally, these single-primed RCA amplicons are produced from ssDNA molecules. However, in some cases, single-primed reactions may also occur for dsDNA, where dsDNA predominates over ssDNA due to the 10^5-fold degree of amplification. In the case of a double-primed RCA reaction, a pair of primers is involved, where one primer works as a complementary and the second primer works as a linear RCA chain. As a result of this, dsDNA fragments shaped like padlock circular form that is produced by this reaction. Figure 6 described the analysis of Kras DNA by RCA amplification, where specific primers bind with targeted Kras DNA by polymerization reaction and displacing the hybridized target DNA that leads to few DNA cycles (Xu et al. 2017). These are visualized as ladder-like bands in the gel. This RCA reaction yields 10^9 copies or more of a circular sequence within 1 h, which is much faster than the conventional PCR or ELISA procedures (Demidov 2002).

In the year 2005, Wang et al. reported SARS-CoV RNA detection by RCA tools. To enhance the sensitivity, RCA assay was performed on serially-diluted nucleic acid templates of interest. The determined LOD was below 100 number of copies. Moreover, in terms of sensitivity, RCA showed more efficiency than RT-PCR which is increased by ~30% (Wang et al. 2005a).

3.1.4 NGS-Based Methodology

Next-generation sequencing (NGS) technologies are an incredible molecular diagnostic tool for detailed genomic research. In the year 2005, first NGS technology was launched in the market. Since then, this technology has been used for standard sequencing applications, like whole genome sequencing, disease detection, and mutation detection (Morozova and Marra 2008). The broad application spectrum

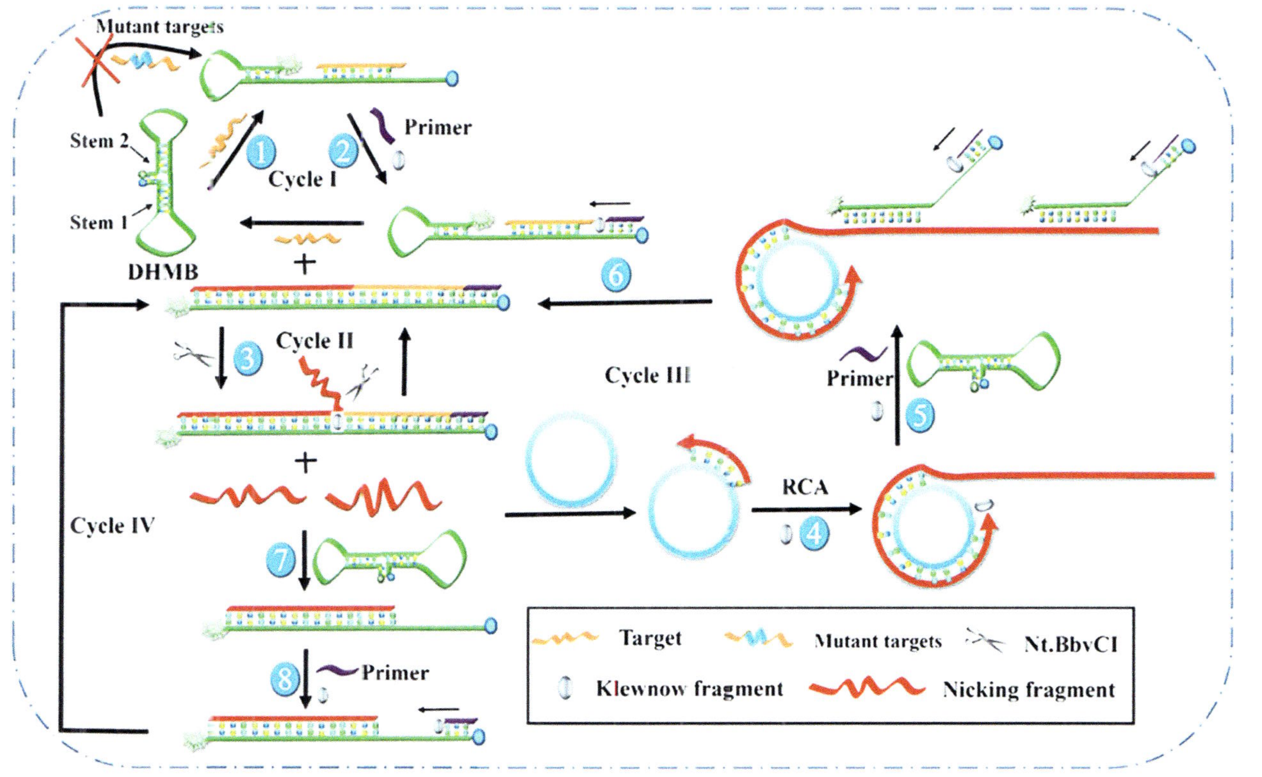

Fig. 5 Schematic representation of RCA amplification and its principle in detail. This figure represents the detection of mutation in the Kras gene by RCA reaction by targeting DNA duplex and polymerization replications by target strand displacement. (This figure is reproduced with permission from Xu et al. (2017))

of NGS in functional genome research has made many things easier, such as gene expression profiling, genome annotation, detection of transcription, non-coding RNA (ncRNA) discovery, nucleic acid sequencing, etc. In Fig. 7, it represents the cDNA sequencing process with reference genomes, where cDNA generates short sequence reads to map their locations and determined the genome distribution. By this sequencing process, the analysis of gene expression of the whole unknown pathogen genome can be established (Marguerat et al. 2008). Different platforms are involved in this technology, for example, Illumina, nanopore DNAsequences, ligation sequencing (SOLiD), single-molecule real-time (SMRT) sequencing, Pyrosequencing, GenapSys sequencing, etc. In terms of high throughput and accuracy, NGS technology has shown much effective molecular processing for better advancement procedures. Progress in DNA sequencing using NGS techniques has been becoming a method of choice because of its faster revolution in sequencing technology (Liu et al. 2012).

A group of researchers performed various kinds of viral genome sequencing using Illumina and nanopore sequencing platform to characterize the virus genome

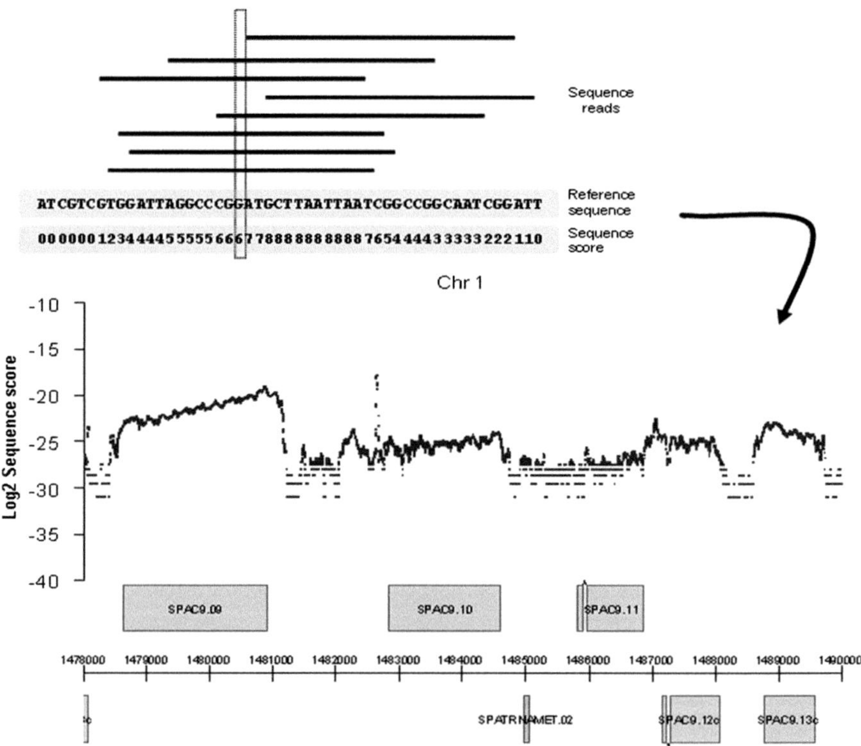

Fig. 7 A generic sequencing process of cDNA that allows genome-wide measurements in transcription levels by RNA-seq. (This figure is reproduced with the permission from Marguerat et al. (2008))

in detail directly from clinical samples (Quick et al. 2017). Recently, one of the studies explained about cross-contamination experiment of COVID-19 with other pathogenic species using this procedure. For that reason, various clinical samples of other human coronaviruses (HCoV-229E, HCoV-NL63, HCoV-oC43, and HCoV-HKU1), including SARS and MERS samples, were considered. RNA was extracted from fluids collected from infected patients and was used as a template to clone. They obtained almost 20,000 viral reads from each clinical specimen, and most of the reads were matched with genome obtained from beta-coronavirus. This result showed an 85% identity with a bat SARS-like CoV genome. They helped to confirm that COVID-19 is the seventh member of the family of coronaviruses that are spreading and infecting humans (Zhu et al. 2020). Another group from Wuhan, China has determined a study on COVID-19 using NGS. Their main aim was to describe the CT findings at different time points through NGS technology. They have experimented on 81 patients, which showed this technology is capable of handling huge numbers of samples. They also compared the results with RT-PCR to confirm the positive tests (Shi et al. 2020).

3.1.5   Microarray-Based Methodology

The new revolutionary analytical tools for diagnosing diseases are biosensors (Mahato et al. 2018a; Kumar et al. 2019a). Along with biosensors, usage of novel nanomaterials bring a new dimension for the disease-detection techniques, and these techniques are being used widely for its more rapid and high-throughput screening at lower expenses (Roy et al. 2019). Many molecular reactions are being applied in the field of biosensors for quick diagnosis process (Mahato et al. 2018b; Kumar et al. 2019b, c; Purohit et al. 2019; Mahato et al. 2020). Among them, the microarray-based methodology is broadly used for disease-diagnosing process (Shi et al. 2003; Zhu et al. 2006). As described by the researchers first, RNA is extracted from clinical samples. Then pairs of primers were designed and incorporated in the master mix reaction. Later, the viral RNA was converted to cDNA with specific probes via the reverse transcription process. Afterward, cDNAs were loaded into each microarray well, and samples were hybridized with targeted cDNAs in a solid phase. This step requires thorough washing to eliminate extra free cDNAs on the surface or in the solution (Long et al. 2004). In the end stage, the viral RNA was detected by the positive signals from the specific probes. Generally, the main origin of DNA/RNA microarray procedure is an approach of high-throughput screening of oligonucleotides with DNA amplification and optical readouts (Fig. 8). This assay is based on convection-controlled hybridization between two NA strands on a paper substrate within a filtration (Rivas et al. 2018). On the other hand, Chen et al. described comprehensive detection and identification of animal and human coronaviruses with microarray assay. They achieved 1000 times more sensitivity compared with normal PCR processes with no cross-reaction between non-targeted strains (Chen et al. 2010).

Due to the rapid mutation of SARS-CoV, Guo et al. determined a microarray assay to detect this virus of 24 single nucleotide polymorphism (SNP) mutations with the gene of SARS-CoV with 100% accuracy (Guo et al. 2014). Therefore, soon

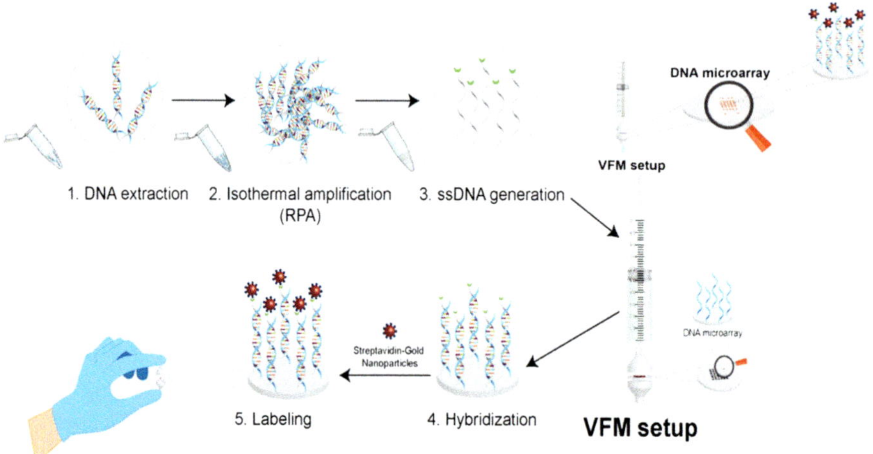

Fig. 8 Schematic representation of a general DNA microarray procedure. (This figure is reproduced with the permission from Rivas et al. (2018))

it is expected that this sudden outbreak can get benefits from this process to diagnose pathogens in a wide range.

4 Challenges and Future Directions

COVID-19 pandemic is rapidly evolving. During this time, everyone is facing lots of challenges and complexity. Especially, for frontliners, like medical staff, the challenges are even more critical. Moreover, handling clinical samples and maintaining those samples properly and contamination-free while processing are difficult. Few challenges/issues raised during this time, we have emphasized below:

(a) Pre-analytical issues: Collection of initial respiratory tract specimens for screening and diagnosis is difficult as the viral loads appeared heavily at the respiratory tracts with the potential to affect the person while collecting samples, if proper measurements are not taken. Samples need to be well maintained followed by proper guidelines provided by the CDC and need extreme handling care (Tang et al. 2020).

(b) Late detection issues: Delayed diagnosis can be crucial for patients, as this may cause lung infection and will increase the acute symptoms, as the virus starts spreading rapidly. Patients who have serious pneumonia and acute respiratory syndrome require early isolations. Since this virus has an incubation period of 2–14 days, testing of this patient samples must be quick. Moreover, if isolation does not occur soon then, unknowingly, this virus can spread to others at a rapid rate.There upon, early detection with proper isolation process can reduce the spreading of this virus (Chan et al. 2004; Li et al. 2020b).

(c) Monitoring patients with severe COVID-19 infection: The diseases have a severe impact on the elderly peoples as shown in the number of maximum fatalities in countries, like Italy, USA, where a big part of the population is old. Therefore, managing patients is a huge problem due to lack of place in the hospitals and fewer numbers of PPE, and numbers of testing kits are less to test patients. Once, anyone is resulted as positive of COVID-19 infections then that particular patient needs to test a few times in a row to regulate the ongoing status of infections. Even, the once infected person is recovered or symptoms got reduced, then also they need to test again to confirm that COVID-19 is negative for them, before they go back to normal routines. Thus, monitoring infected patients continuously is essential until they are recovered. This whole process is a hectic issue.

(d) Safety measures for sample processing and testing: Whether a person who is collecting samples and a person who is performing laboratory tests, everybody needs to follow the safety protocols provided by the CDC. Processing of respiratory samples should be maintained in a class II biological safety guidelines/biosafety level three (BSL3) (Chu et al. 2020).

(e) Analytical issues: Most of the critical challenges arose while testing or diagnosing the samples. Sometimes confusion arises for assay selection, since so many assays are available currently, and those are very competitive, capable of early diagnosis. During experiments, there are few chances to occur false-positive results by RT-PCR or RT-LAMP. Although these procedures give rapid solutions, rarely it also shows cross-contamination issues. So, specificity is another major issue for analytical analysis.

(f) Post analytical issues: After performing various experiments, the problem comes while analyzing a huge amount of data. Accurate interpretation of molecular results is extremely crucial for precise disease diagnosis.

To date, significant determinations have been made to improve the detection of coronaviruses, and various improved, and new approaches have been employed. Advancement of the amplification process through techniques, like PCR, LAMP, and NGS, will continue to lead the rapid diagnostics. Reduction in time and cost of the usage of these molecular processes will result in easy availability for all patients. Despite all, using advanced technology therapeutic solutions and vaccines will come soon in the future. We believe that continuous efforts from researchers will establish a more outstanding and capable diagnosis method to prevail in the viral outbreak.

5   Conclusion

The ongoing outbreak of COVID-19 infections worldwide has affirmed the attention of laboratory-based diagnosis procedures to restrain the spread. With the progress of emerging molecular technologies, researchers are playing the major role to develop quick remedies for this difficult situation. The very essential key to react to these pandemics is to monitor the symptoms, since other mechanisms were unknown.

Further, molecular diagnosis processes are playing a critical role in early detection of this virus in infected patients and the source of infection. The capacity of molecular diagnosis tools of coronaviruses reduces the detection time and the risks of spreading this virus as well. All these technologies are simple and straightforward and follow standard operating procedures. Due to its simplicity with fewer reagents and few instrumentations, these are now preferred and widely used for disease diagnosis. This chapter has summarized the current molecular processes that are available for COVID-19 and other coronaviruses. We have also discussed current researches going on for diagnosing the novel coronavirus by RT-PCR, RT-LAMP, RCA, and NGS technologies. We also explained sample processing that is followed by the CDC. Besides, we have drawn a few points about the challenges that arise due to this pandemic. However, many unknown questions have remained unsolved and that more studies are needed.

References

Adachi D, Johnson G, Draker R, Ayers M, Mazzulli T, Talbot PJ, Tellier R (2004) Comprehensive detection and identification of human coronaviruses, including the SARS-associated coronavirus, with a single RT-PCR assay. J Virol Methods 122(1):29–36

Azam NFN, Roy S, Lim SA, Uddin Ahmed M (2018) Meat species identification using DNA-luminol interaction and their slow diffusion onto the biochip surface. Food Chem 248:29–36

Balboni A, Gallina L, Palladini A, Prosperi S, Battilani M (2012) A real-time PCR assay for bat SARS-like coronavirus detection and its application to Italian greater horseshoe bat Faecal sample surveys. Sci World J 2012:989514

Bloomfield MG, Balm MND, Blackmore TK (2015) Molecular testing for viral and bacterial enteric pathogens: gold standard for viruses, but don't let culture go just yet? Pathology 47(3):227–233

Chan PKS, To W-K, Ng K-C, Lam RKY, Ng T-K, Chan RCW, Wu A, Yu W-C, Lee N, Hui DSC, Lai S-T, Hon EKL, Li C-K, Sung JJY, Tam JS (2004) Laboratory diagnosis of SARS. Emerg Infect Dis 10(5):825–831

Chan JF-W, Yip CC-Y, To K K-W, Tang TH-C, Wong SC-Y, Leung K-H, Fung AY-F, Ng AC-K, Zou Z, Tsoi H-W, Choi GK-Y, Tam AR, Cheng VC-C, Chan K-H, Tsang OT-Y, Yuen K-Y (2020) Improved molecular diagnosis of COVID-19 by the novel, highly sensitive and specific COVID-19-RdRp/Hel real-time reverse transcription-PCR assay validated in vitro and with clinical specimens. J Clin Microbiol 58(5):e00310–e00320

Chapin SC, Doyle PS (2011) Ultrasensitive multiplexed MicroRNA quantification on encoded gel microparticles using rolling circle amplification. Anal Chem 83(18):7179–7185

Chen Q, Li J, Deng Z, Xiong W, Wang Q, Hu Y (2010) Comprehensive detection and identification of seven animal coronaviruses and human respiratory coronavirus 229E with a microarray hybridization assay. Intervirology 53(2):95–104

Chu DKW, Poon LLM, Gomaa MM, Shehata MM, Perera RAPM, Abu Zeid D, El Rifay AS, Siu LY, Guan Y, Webby RJ, Ali MA, Peiris M, Kayali G (2014) MERS coronaviruses in dromedary camels, Egypt. Emerg Infect Dis 20(6):1049–1053

Chu DKW, Pan Y, Cheng SMS, Hui KPY, Krishnan P, Liu Y, Ng DYM, Wan CKC, Yang P, Wang Q, Peiris M, Poon LLM (2020) Molecular diagnosis of a novel coronavirus (2019-nCoV) causing an outbreak of pneumonia. Clin Chem 66(4):549–555

Corman V, Eckerle I, Bleicker T, Zaki A, Landt O, Eschbach-Bludau M, van Boheemen S, Gopal R, Ballhause M, Bestebroer T, Muth D, Müller M, Drexler J-F, Zambon M,

Osterhaus A, Fouchier R, Drosten C (2012) Detection of a novel human coronavirus by real-time reverse-transcription polymerase chain reaction. Eur Secur 17(39)

Corman VM, Landt O, Kaiser M, Molenkamp R, Meijer A, Chu DK, Bleicker T, Brünink S, Schneider J, Schmidt ML, Mulders DG, Haagmans BL, van der Veer B, van den Brink S, Wijsman L, Goderski G, Romette J-L, Ellis J, Zambon M, Peiris M, Goossens H, Reusken C, Koopmans MP, Drosten C (2020) Detection of 2019 novel coronavirus (2019-nCoV) by real-time RT-PCR. Eur Secur 25(3):2000045

Demidov VV (2002) Rolling-circle amplification in DNA diagnostics: the power of simplicity. Expert Rev Mol Diagn 2(6):542–548

Drosten C, Meyer B, Müller MA, Corman VM, Al-Masri M, Hossain R, Madani H, Sieberg A, Bosch BJ, Lattwein E, Alhakeem RF, Assiri AM, Hajomar W, Albarrak AM, Al-Tawfiq JA, Zumla AI, Memish ZA (2014) Transmission of MERS-coronavirus in household contacts. N Engl J Med 371(9):828–835

Enosawa M, Kageyama S, Sawai K, Watanabe K, Notomi T, Onoe S, Mori Y, Yokomizo Y (2003) Use of loop-mediated isothermal amplification of the IS900 sequence for rapid detection of cultured mycobacterium avium subsp. paratuberculosis. J Clin Microbiol 41(9):4359

Giuseppe L, Ana-Maria S, Mario P (2020) Potential preanalytical and analytical vulnerabilities in the laboratory diagnosis of coronavirus disease 2019 (COVID-19). Clin Chem Lab Med (CCLM):20200285

Gu J, Han B, Wang J (2020) COVID-19: gastrointestinal manifestations and potential fecal oral transmission. Gastroenterology 158(6):1518–1519

Guo X, Geng P, Wang Q, Cao B, Liu B (2014) Development of a single nucleotide polymorphism DNA microarray for the detection and genotyping of the SARS coronavirus. J Microbiol Biotechnol 24(10):1445–1454

Hadjinicolaou AV, Farcas GA, Demetriou VL, Mazzulli T, Poutanen SM, Willey BM, Low DE, Butany J, Asa SL, Kain KC, Kostrikis LG (2011) Development of a molecular-beacon-based multi-allelic real-time RT-PCR assay for the detection of human coronavirus causing severe acute respiratory syndrome (SARS-CoV): a general methodology for detecting rapidly mutating viruses. Arch Virol 156(4):671–680

Huang P, Wang H, Cao Z, Jin H, Chi H, Zhao J, Yu B, Yan F, Hu X, Wu F, Jiao C, Hou P, Xu S, Zhao Y, Feng N, Wang J, Sun W, Wang T, Gao Y, Yang S, Xia X (2018) A rapid and specific assay for the detection of MERS-CoV. Front Microbiol 9(1101)

Ibrahim IM, Abdelmalek DH, Elshahat E, Elfiky AA (2020) COVID-19 spike-host cell receptor GRP78 binding site prediction. J Infect 80(5):554–562

Jiang SS, Chen T-C, Yang J-Y, Hsiung CA, Su I-J, Liu Y-L, Chen P-C, Juang J-L (2004) Sensitive and quantitative detection of severe acute respiratory syndrome coronavirus infection by real-time nested polymerase chain reaction. Clin Infect Dis 38(2):293–296

Jiang M, Pan W, Arastehfar A, Fang W, Ling L, Fang H, Farnaz Daneshnia F, Yu J, Liao W, Pei H, Li X, Lass-Florl C (2020) Development and validation of a rapid single-step reverse transcriptase loop-mediated isothermal amplification (RT-LAMP) system potentially to be used for reliable and high-throughput screening of COVID-19. medRxiv: 2020.2003.2015.20036376

Jung IY, You JB, Choi BR, Kim JS, Lee HK, Jang B, Jeong HS, Lee K, Im SG, Lee H (2016) A highly sensitive molecular detection platform for robust and facile diagnosis of Middle East respiratory syndrome (MERS) Corona virus. Adv Healthc Mater 5(17):2168–2173

Jung YJ, Park G-S, Moon JH, Ku K, Beak S-H, Kim S, Park EC, Park D, Lee J-H, Byeon CW, Lee JJ, Maeng J-S, Kim SJ, Kim SI, Kim BT, Lee MJ, Kim HG (2020) Comparative analysis of primer-probe sets for the laboratory confirmation of SARS-CoV-2. bioRxiv: 2020.2002.2025.964775

Klein D (2002) Quantification using real-time PCR technology: applications and limitations. Trends Mol Med 8(6):257–260

Kumar A, Purohit B, Mahato K, Chandra P (2019a) Chapter 11 advance engineered nanomaterials in point-of-care immunosensing for biomedical diagnostics. Immunosensors, The Royal Society of Chemistry, pp 238–266

Kumar A, Purohit B, Maurya PK, Pandey LM, Chandra P (2019b) Engineered nanomaterial assisted signal-amplification strategies for enhancing analytical performance of electrochemical biosensors. Electroanalysis 31(9):1615–1629

Kumar A, Roy S, Srivastava A, Naikwade MM, Purohit B, Mahato K, Naidu VGM, Chandra P (2019c) Chapter 10 – Nanotherapeutics: a novel and powerful approach in modern healthcare system. In: Maurya PK, Singh S (eds) Nanotechnology in modern animal biotechnology. Elsevier, pp 149–161

Lamb LE, Bartolone SN, Ward E, Chancellor MB (2020) Rapid detection of novel coronavirus (COVID-19) by reverse transcription-loop-mediated isothermal amplification. medRxiv: 2020.2002.2019.20025155

Lan L, Xu D, Ye G, Xia C, Wang S, Li Y, Xu H (2020) Positive RT-PCR test results in patients recovered from COVID-19. JAMA 323(15):1502–1503

Lau YL, Meganathan P, Sonaimuthu P, Thiruvengadam G, Nissapatorn V, Chen Y (2010) Specific, sensitive, and rapid diagnosis of active toxoplasmosis by a loop-mediated isothermal amplification method using blood samples from patients. J Clin Microbiol 48(10):3698

Li Y, Fan P, Zhou S, Zhang L (2017) Loop-mediated isothermal amplification (LAMP): a novel rapid detection platform for pathogens. Microb Pathog 107:54–61

Li D, Wang D, Dong J, Wang N, Huang H, Xu H, Xia C (2020a) False-negative results of real-time reverse-transcriptase polymerase chain reaction for severe acute respiratory syndrome corona-virus 2: role of deep-learning-based CT diagnosis and insights from two cases. Korean J Radiol 21(4):505–508

Li Q, Guan X, Wu P, Wang X, Zhou L, Tong Y, Ren R, Leung KSM, Lau EHY, Wong JY, Xing X, Xiang N, Wu Y, Li C, Chen Q, Li D, Liu T, Zhao J, Liu M, Tu W, Chen C, Jin L, Yang R, Wang Q, Zhou S, Wang R, Liu H, Luo Y, Liu Y, Shao G, Li H, Tao Z, Yang Y, Deng Z, Liu B, Ma Z, Zhang Y, Shi G, Lam TTY, Wu JT, Gao GF, Cowling BJ, Yang B, Leung GM, Feng Z (2020b) Early transmission dynamics in Wuhan, China, of novel coronavirus–infected pneumonia. N Engl J Med 382(13):1199–1207

Li X, Zai J, Wang X, Li Y (2020c) Potential of large "first generation" human-to-human transmission of 2019-nCoV. J Med Virol 92(4):448–454

Liu L, Li Y, Li S, Hu N, He Y, Pong R, Lin D, Lu L, Law M (2012) Comparison of next-generation sequencing systems. J Biomed Biotechnol 2012:251364

Liu Y, Gayle AA, Wilder-Smith A, Rocklöv J (2020) The reproductive number of COVID-19 is higher compared to SARS coronavirus. J Travel Med 27(2)

Long W-H, Xiao H-S, Gu X-M, Zhang Q-H, Yang H-J, Zhao G-P, Liu J-H (2004) A universal microarray for detection of SARS coronavirus. J Virol Methods 121(1):57–63

Lu R, Wu X, Wan Z, Li Y, Zuo L, Qin J, Jin X, Zhang C (2020) Development of a novel reverse transcription loop-mediated isothermal amplification method for rapid detection of SARS-CoV-2. Virologica Sinica

Mahato K, Kumar A, Maurya PK, Chandra P (2018a) Shifting paradigm of cancer diagnoses in clinically relevant samples based on miniaturized electrochemical nanobiosensors and microfluidic devices. Biosens Bioelectron 100:411–428

Mahato K, Kumar S, Srivastava A, Maurya PK, Singh R, Chandra P (2018b) Chapter 14 – electrochemical Immunosensors: fundamentals and applications in clinical diagnostics. In: Vashist SK, Luong JHT (eds) Handbook of immunoassay technologies. Academic, pp 359–414

Mahato K, Purohit B, Kumar A, Chandra P (2020) Clinically comparable impedimetric immunosensor for serum alkaline phosphatase detection based on electrochemically engineered Au-nano-Dendroids and graphene oxide nanocomposite. Biosens Bioelectron 148:111815

Marguerat S, Wilhelm BT, Bähler J (2008) Next-generation sequencing: applications beyond genomes. Biochem Soc Trans 36(5):1091–1096

Morozova O, Marra MA (2008) Applications of next-generation sequencing technologies in functional genomics. Genomics 92(5):255–264

Mothershed EA, Whitney AM (2006) Nucleic acid-based methods for the detection of bacterial pathogens: present and future considerations for the clinical laboratory. Clin Chim Acta 363 (1):206–220

Nagamine K, Watanabe K, Ohtsuka K, Hase T, Notomi T (2001) Loop-mediated isothermal amplification reaction using a nondenatured template. Clin Chem 47(9):1742–1743

Nagaraj T, Vasanth JP, Desai A, Kamat A, Madhusudana SN, Ravi V (2006) Ante mortem diagnosis of human rabies using saliva samples: comparison of real time and conventional RT-PCR techniques. J Clin Virol 36(1):17–23

Ni W, Le Guiner C, Moullier P, Snyder RO (2012) Development and utility of an internal threshold control (ITC) real-time PCR assay for exogenous DNA detection. PLoS One 7(5):e36461–e36461

Nilsson M, Malmgren H, Samiotaki M, Kwiatkowski M, Chowdhary BP, Landegren U (1994) Padlock probes: circularizing oligonucleotides for localized DNA detection. Science 265 (5181):2085

Notomi T, Okayama H, Masubuchi H, Yonekawa T, Watanabe K, Amino N, Hase T (2000) Loop-mediated isothermal amplification of DNA. Nucleic Acids Res 28(12):e63–e63

Notomi T, Mori Y, Tomita N, Kanda H (2015) Loop-mediated isothermal amplification (LAMP): principle, features, and future prospects. J Microbiol 53(1):1–5

Pan Y, Li X, Yang G, Fan J, Tang Y, Zhao J, Long X, Guo S, Zhao Z, Liu Y, Hu H, Xue H, Li Y (2020) Serological immunochromatographic approach in diagnosis with SARS-CoV-2 infected COVID-19 patients. J Infect 81:e28–e32

Park G-S, Ku K, Baek S-H, Kim S-J, Kim SI, Kim B-T, Maeng J-S (2020) Development of reverse transcription loop-mediated isothermal amplification assays targeting severe acute respiratory syndrome coronavirus 2. J Mol Diagn 22:729–735

Poon LLM, Leung CSW, Tashiro M, Chan KH, Wong BWY, Yuen KY, Guan Y, Peiris JSM (2004) Rapid detection of the severe acute respiratory syndrome (SARS) coronavirus by a loop-mediated isothermal amplification assay. Clin Chem 50(6):1050–1052

Purohit B, Mahato K, Kumar A, Chandra P (2019) Sputtering enhanced peroxidase like activity of a dendritic nanochip for amperometric determination of hydrogen peroxide in blood samples. Microchim Acta 186(9):658

Pyrc K, Milewska A, Potempa J (2011) Development of loop-mediated isothermal amplification assay for detection of human coronavirus-NL63. J Virol Methods 175(1):133–136

Quick J, Grubaugh ND, Pullan ST, Claro IM, Smith AD, Gangavarapu K, Oliveira G, Robles-Sikisaka R, Rogers TF, Beutler NA, Burton DR, Lewis-Ximenez LL, de Jesus JG, Giovanetti M, Hill SC, Black A, Bedford T, Carroll MW, Nunes M, Alcantara LC, Sabino EC, Baylis SA, Faria NR, Loose M, Simpson JT, Pybus OG, Andersen KG, Loman NJ (2017) Multiplex PCR method for MinION and Illumina sequencing of Zika and other virus genomes directly from clinical samples. Nat Protoc 12(6):1261–1276

Riou J, Althaus CL (2020) Pattern of early human-to-human transmission of Wuhan 2019 novel coronavirus (2019-nCoV), December 2019 to January 2020. Eur Secur 25(4):2000058

Rivas L, Reuterswärd P, Rasti R, Herrmann B, Mårtensson A, Alfvén T, Gantelius J, Andersson-Svahn H (2018) A vertical flow paper-microarray assay with isothermal DNA amplification for detection of Neisseria meningitidis. Talanta 183:192–200

Roy S, Rahman IA, Ahmed MU (2016a) Paper-based rapid detection of pork and chicken using LAMP–magnetic bead aggregates. Anal Methods 8(11):2391–2399

Roy S, Rahman IA, Santos JH, Ahmed MU (2016b) Meat species identification using DNA-redox electrostatic interactions and non-specific adsorption on graphene biochips. Food Control 61:70–78

Roy S, Wei SX, Ying JLZ, Satavieh M, Ahmed MU (2016c) A novel, sensitive and label-free loop-mediated isothermal amplification detection method for nucleic acids using luminophore dyes. Biosens Bioelectron 86:346–352

Roy S, Hossain MM, Safavieh M, Lubis HN, Zourob M, Ahmed MU (2017a) Chapter 16 isothermal DNA amplification strategies for food biosensors. In: Food biosensors. The Royal Society of Chemistry, Cambridge, pp 367–392

Roy S, Mohd-Naim NF, Safavieh M, Ahmed MU (2017b) Colorimetric nucleic acid detection on paper microchip using loop mediated isothermal amplification and crystal violet dye. ACS Sens 2(11):1713–1720

Roy S, Malode SJ, Shetti NP, Chandra P (2019) Modernization of biosensing strategies for the development of lab-on-chip integrated systems. Bioelectrochemical Interface Engineering, pp 325–342

Setianingsih TY, Wiyatno A, Hartono TS, Hindawati E, Rosamarlina A, Dewantari K, Myint KS, Lisdawati V, Safari D (2019) Detection of multiple viral sequences in the respiratory tract samples of suspected Middle East respiratory syndrome coronavirus patients in Jakarta, Indonesia 2015–2016. Int J Infect Dis 86:102–107

Sheet OH, Grabowski NT, Klein G, Abdulmawjood A (2016) Development and validation of a loop mediated isothermal amplification (LAMP) assay for the detection of Staphylococcus aureus in bovine mastitis milk samples. Mol Cell Probes 30(5):320–325

Shen M, Zhou Y, Ye J, Abdullah Al-maskri AA, Kang Y, Zeng S, Cai S (2020) Recent advances and perspectives of nucleic acid detection for coronavirus. J Pharm Anal 10(2):97–101

Shi R, Ma W, Wu Q, Zhang B, Song Y, Guo Q, Xiao W, Wang Y, Zheng W (2003) Design and application of 60mer oligonucleotide microarray in SARS coronavirus detection. Chin Sci Bull 48(12):1165–1169

Shi H, Han X, Jiang N, Cao Y, Alwalid O, Gu J, Fan Y, Zheng C (2020) Radiological findings from 81 patients with COVID-19 pneumonia in Wuhan, China: a descriptive study. Lancet Infect Dis 20(4):425–434

Shirato K, Yano T, Senba S, Akachi S, Kobayashi T, Nishinaka T, Notomi T, Matsuyama S (2014) Detection of Middle East respiratory syndrome coronavirus using reverse transcription loop-mediated isothermal amplification (RT-LAMP). Virol J 11(1):139

Tahamtan A, Ardebili A (2020) Real-time RT-PCR in COVID-19 detection: issues affecting the results. Expert Rev Mol Diagn 20(5):453–454

Tang Y-W, Schmitz JE, Persing DH, Stratton CW (2020) The laboratory diagnosis of COVID-19 infection: current issues and challenges. J Clin Microbiol JCM 58:00512–00520

Udugama B, Kadhiresan P, Kozlowski HN, Malekjahani A, Osborne M, Li VYC, Chen H, Mubareka S, Gubbay JB, Chan WCW (2020) Diagnosing COVID-19: the disease and tools for detection. ACS Nano 14(4):3822–3835

Ushikubo H (2004) Principle of LAMP method–a simple and rapid gene amplification method. Uirusu 54(1):107–112

van Elden LJR, Anton AM, van Alphen F, Hendriksen KAW, Hoepelman AIM, van Kraaij MGJ, Oosterheert J-J, Schipper P, Schuurman R, Nijhuis M (2004) Frequent detection of human coronaviruses in clinical specimens from patients with respiratory tract infection by use of a novel real-time reverse-transcriptase polymerase chain reaction. J Infect Dis 189(4):652–657

Velavan TP, Meyer CG (2020) The COVID-19 epidemic. Trop Med Int Health 25(3):278–280

Wang B, Potter SJ, Lin Y, Cunningham AL, Dwyer DE, Su Y, Ma X, Hou Y, Saksena NK (2005a) Rapid and sensitive detection of severe acute respiratory syndrome coronavirus by rolling circle amplification. J Clin Microbiol 43(5):2339

Wang W-K, Fang C-T, Chen H-L, Yang C-F, Chen Y-C, Chen M-L, Chen S-Y, Yang J-Y, Lin J-H, Yang P-C, Chang S-C (2005b) Detection of severe acute respiratory syndrome coronavirus RNA in plasma during the course of infection. J Clin Microbiol 43(2):962

Weiss SR, Navas-Martin S (2005) Coronavirus pathogenesis and the emerging pathogen severe acute respiratory syndrome coronavirus. Am Soc Microbiol J 69(4):635–664

Xu H, Wu D, Jiang Y, Zhang R, Wu Q, Liu Y, Li F, Wu Z-S (2017) Loopback rolling circle amplification for ultrasensitive detection of Kras gene. Talanta 164:511–517

Xu M, Ye J, Yang D, Abdullah Al-maskri AA, Hu H, Jung C, Cai S, Zeng S (2019) Ultrasensitive detection of miRNA via one-step rolling circle-quantitative PCR (RC-qPCR). Anal Chim Acta 1077:208–215

Yip SP, To SST, Leung PHM, Cheung TS, Cheng PKC, Lim WWL (2005) Use of dual TaqMan probes to increase the sensitivity of 1-step quantitative reverse transcription-PCR: application to the detection of SARS coronavirus. Clin Chem 51(10):1885–1888

Yu X-F, Pan J-C, Ye R, Xiang H-Q, Kou Y, Huang Z-C (2008) Preparation of armored RNA as a control for multiplex real-time reverse transcription-PCR detection of influenza virus and severe acute respiratory syndrome coronavirus. J Clin Microbiol 46(3):837

Yu L, Wu S, Hao X, Li X, Liu X, Ye S, Han H, Dong X, Li X, Li J, Liu J, Liu N, Zhang W, Pelechano V, Chen W-H, X Yin (2020) Rapid colorimetric detection of COVID-19 coronavirus using a reverse tran-scriptional loop-mediated isothermal amplification (RT-LAMP) diagnostic plat-form: iLACO. medRxiv: 2020.2002.2020.20025874

Zhang Y, Odiwuor N, Xiong J, Sun L, Nyaruaba RO, Wei H, Tanner NA (2020) Rapid molecular detection of SARS-CoV-2 (COVID-19) virus RNA using colorimetric LAMP. medRxiv: 2020.2002.2026.20028373

Zhou P, Yang X-L, Wang X-G, Hu B, Zhang L, Zhang W, Si H-R, Zhu Y, Li B, Huang C-L, Chen H-D, Chen J, Luo Y, Guo H, Jiang R-D, Liu M-Q, Chen Y, Shen X-R, Wang X, Zheng X-S, Zhao K, Chen Q-J, Deng F, Liu L-L, Yan B, Zhan F-X, Wang Y-Y, Xiao G-F, Shi Z-L (2020a) A pneumonia outbreak associated with a new coronavirus of probable bat origin. Nature 579 (7798):270–273

Zhou T, Liu Q, Yang Z, Liao J, Yang K, Bai W, Lu X, Zhang W (2020b) Preliminary prediction of the basic reproduction number of the Wuhan novel coronavirus 2019-nCoV. J Evid Based Med 13(1):3–7

Zhu H, Hu S, Jona G, Zhu X, Kreiswirth N, Willey BM, Mazzulli T, Liu G, Song Q, Chen P, Cameron M, Tyler A, Wang J, Wen J, Chen W, Compton S, Snyder M (2006) Severe acute respiratory syndrome diagnostics using a coronavirus protein microarray. Proc Natl Acad Sci U S A 103(11):4011

Zhu N, Zhang D, Wang W, Li X, Yang B, Song J, Zhao X, Huang B, Shi W, Lu R, Niu P, Zhan F, Ma X, Wang D, Xu W, Wu G, Gao GF, Tan W (2020) A novel coronavirus from patients with pneumonia in China, 2019. N Engl J Med 382(8):727–733

Futuristic Technologies for Advanced Detection, Prevention, and Control of COVID-19

Sarvodaya Tripathy, Russell Kabir, S. M. Yasir Arafat, and Shailendra K. Saxena ⓘ

Abstract

The COVID-19 pandemic has been devastatingly taking tolls on almost every aspect of human life across the globe. Although a good number of studies have been coming out, some of the characteristics of the novel coronavirus are yet unknown. Early diagnosis and virus containment are the vital steps to treat, control, and prevent the diseases. However, the majority of the infected patients may be asymptomatic while they are spreading the infection. Therefore, at the community level, primary prevention is fundamentally challenging. A diagnostic test that is easily available, widely acceptable, incurring a low cost, and giving quick response could be an ideal one. Nevertheless, the identification of asymptomatic carriers is the core challenge to prevent spreading. Different countries have been experimenting with different prospecting diagnostic tools where

Authors Sarvodaya Tripathy, Shailendra K. Saxena have equally contributed to this chapter as first author.

S. Tripathy (✉)
Department of Microbiology, Maharaja Krishna Chandra Gajapati Medical College & Hospital, Brahmapur, India

R. Kabir
School of Allied Health, Faculty of Health, Education, Medicine, and Social Care, Anglia Ruskin University, Chelmsford, UK

S. M. Y. Arafat
Department of Psychiatry, Enam Medical College and Hospital, Dhaka, Bangladesh

S. K. Saxena
Centre for Advanced Research (CFAR), Faculty of Medicine, King George's Medical University (KGMU), Lucknow, India
e-mail. shailen@kgmcindia.edu

© The Editor(s) (if applicable) and The Author(s), under exclusive license to Springer Nature Singapore Pte Ltd. 2020
P. Chandra, S. Roy (eds.), *Diagnostic Strategies for COVID-19 and other Coronaviruses*, Medical Virology: From Pathogenesis to Disease Control, https://doi.org/10.1007/978-981-15-6006-4_8

technology has been playing a crucial role. This chapter discusses the prospective technological prospects for the detection, prevention, and control of COVID-19.

Keywords

Futuristic technologies · COVID-19 · Pandemic · Detection · Prevention · Control

1   Introduction

COVID-19, a pandemic declared by the World Health Organization (WHO), caught the world off-guard and has changed human behavior in almost every aspect. It has not only brought remarkable changes in our social life but also, with rising unemployment and de-growing world gross domestic product (GDP), it has yielded us significant economic loss. According to the International Monetary Fund April 2020 outlook, global GDP is expected to decline by 3% in 2020, a result that is worse than the fallout of the global financial crisis (2008–2009). While countries across the globe undertook various measures, including lockdown, social distancing, and restrictions on international travels, to curtail the spread of the disease, these measures have slowed down economic activity at large, pushing policymakers now to choose between the two evils – rising infections and economic slowdown. With no vaccines in the near-term commercialization stage, countries are brainstorming on the most ideal strategy to open-up their respective economies. At this crossroad, a nation's *testing efficacy* becomes an important determinant of its readiness to resume economic activity and bring life back to normal. It doesn't need too much of thought to realize that the result, optimal or otherwise, of a testing ecosystem for COVID-19-like situation hinges on three core factors – the scope of testing, testing infrastructure, and efficiency of testing methods employed. Scope of testing in a country could range from *reactive policy*, which advocates testing limited suspected cases (like people with travel history, contact history, or symptomatic individuals) to *proactive policy*, in which governments and organizations go that extra mile to determine potential risk areas and establish ways and means to conduct tests proactively. An ideal testing infrastructure could broadly be defined by seamless availability of test kits, an adequate number of testing centers, the presence of trained testing professionals, and a harmonious supply chain. The efficiency of testing methods could be achieved by relying on tests that produce accurate results and by establishing standard practices on the choice of tests depending on the testing population and condition on the ground (Fig. 1). Hence, the ability to effectively reach and screen a higher proportion of the population, building robust testing infrastructure, and standardizing sound testing approaches are the essential pillars of an ideal testing ecosystem. While these three proponents have to work seamlessly to maintain the testing standards, any intervention that can boost the quality of one or more of these proponents will help to move a country one-step closer towards its ideal target.

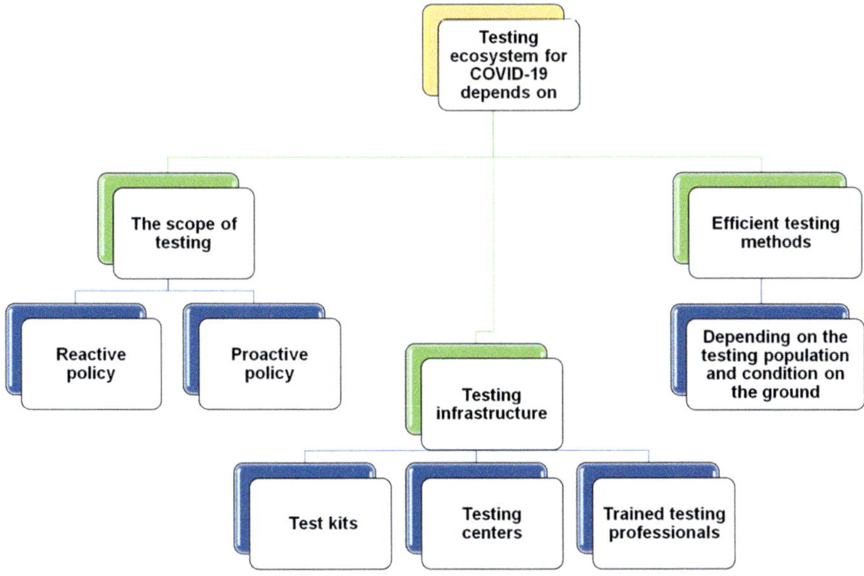

Fig. 1 Testing ecosystem for COVID-19

2   Challenges with Current Diagnostic Techniques

The last three coronavirus outbreaks at the beginning of the twenty-first century stressed the significant warrant of immediate, accurate, and rapid availability of a diagnostic tool to control emerging and re-emerging infections (Younes et al. 2020). SARS-CoV-2 viruses are closely related to bat coronavirus. COVID-19 started as a zoonotic disease with limited human-to-human transmission. It has been shown to be transmitted via contaminated droplets generated due to talking, coughing, sneezing, aerosols generated from the clinical procedure, and handshaking or by contact with contaminated surfaces (Chan et al. 2020; Sabino-Silva et al. 2020). With SARS-CoV-2 infection, patients may remain asymptomatic for a while, and transmission can occur even in this asymptomatic phase (Chan et al. 2020; Sabino-Silva et al. 2020). A recent report suggested that SARS CoV 2 nucleic acid is found in COVID 19 positive patients' urine and feces, indicating that SARS-CoV-2 can transmit through the digestive tract through the feco–oral route (Yang and Wang 2020). Effective pandemic management depends on prompt identification and containment of the infection by strict surveillance, isolation mechanism, and early diagnosis (Soraya and Ulhaq 2020; Younes et al. 2020). Some of the characteristics of the novel coronavirus are yet unknown and limited information on its pathogenesis (Bhusare et al. 2020; Sabino-Silva et al. 2020). Though many therapeutic options of antiviral, steroids, ayurvedic, allopathic, and traditional Chinese medicines were found ineffective against infection, currently no operative and special treatment has been identified by the scientists (Bhusare et al. 2020).

When the virus was beginning to spread from the Wuhan City, China to different corners of the world due to global travel, enough testing capacity was not available, and different countries opted for different diagnostic approaches depending upon availability (Mahapatra and Chandra 2020; Padhi et al. 2020). However, testing kits were available within three days in China and within six weeks in some other countries, and most of the countries initially provided the testing kits to the people who were severely ill. The disadvantage of using testing kits is that they might not be suitable for confirmation of the test result. This may delay the process, and the infection may spread to a larger population in the meantime (Ravichandran et al. 2020).

Real-time reverse-transcription-polymerase-chain-reaction (RT-PCR) assay for COVID-19 has been developed and used in clinics of China, but due to a high false-negative rate and unavailability of RT-PCR, prompt diagnosis of the patient during the initial stage of the outbreak was restricted (Zu et al. 2020). For better result in RT-PCR-based method, good number of viral copies should be present in the sample. There is a high possibility of obtaining false results if the viral replication time window is not gauged, and improperly collected sample may not be useful at all (Bhusare et al. 2020). In low-resource settings, insufficient laboratory equipment and lack of skilled professionals are the main barriers to run molecular diagnostic testing. In addition to it, long turnaround time leads to delay in confirmation of the infected cases, delays in clinical judgment, and increased infection rates (Soraya and Ulhaq 2020; Zoabi and Shomron 2020). The use of bronchoscopy is not a viable diagnostic method, as the aerosol generated may infect both patients and healthcare providers (Pascarella et al. 2020).

Novel coronavirus infection is diagnosed, based on patient's history, clinical features, imaging characteristics, and laboratory tests. Among them, chest CT scan examination plays a vital role in the initial diagnosis of novel coronavirus (Xu et al. 2020). Chest radiography is a useful tool to find an early stage of the infection, but it was not recommended as the first line of diagnostic modality for COVID-19 in China (Zu et al. 2020). The recent report also suggested that a patient with RT-PCR-positive results can have a normal CT scan during the time of hospital admission. However, changes may appear after two to three days of admission (Shuja et al. 2020). Molecular diagnostic tests are more accurate than CT examination and serological tests, as they can identify the specific antigen of SARS-CoV-2 virus. The development of a molecular diagnostic test relies upon understanding the proteomic and genomic configuration of the virus. The proteomic and genomic configuration of SARS-CoV-2 has recently been identified; nevertheless, the host response to the virus is still undiscovered (Younes et al. 2020).

Alizargar et al. (2020) recommended using saliva samples as a safer and convenient way of diagnosing COVID-19 (Alizargar et al. 2020). Tozzi et al. (2020) stated that rectal swab could be an effective way to identify COVID-19-positive cases. Symptomatic individuals with negative nasopharyngeal swab test can be tested with rectal swab (Tozzi et al. 2020). The testing of COVID-19 will vary among countries,

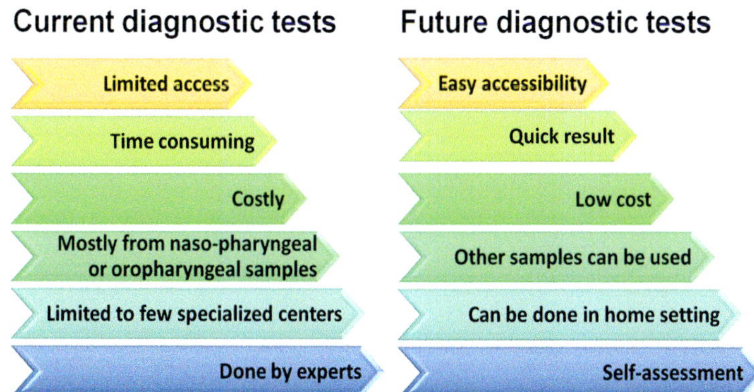

Fig. 2 Current and future diagnostic tests for COVID-19

but testing facilities and tools can be developed basing on experiences from other countries. An ideal diagnostic test should be highly sensitive, specific, easily available at an affordable price, and should be able to give quick results. Considering the current scenario, the diagnostic tests used are unable to meet the above characteristics. Hence, there is a need to use innovations in technology to meet future demands (Fig. 2).

3 Technology and Newer COVID-19 Diagnostics

Though we believe that the broader solution to a nation's testing problem is effective implementation of the healthcare-government-technology (HGT) model, technology has a disproportionate role to play here. Technology can be leveraged not only by using it as an enabler, which connects healthcare and government on a platform but also by providing solutions to the real-world problems, i.e., helping build a true cohesive HGT model, thereby bettering the quality of testing ecosystem. To better appreciate the potential of technological interventions, we have divided this section into three sub-sections. The first sub-section talks about current experimentation underplay and comments on the effectiveness of a few nuanced technological measures undertaken by select countries to tackle COVID-19. This will follow-up with a review of some past epidemics and learnings from them from a technological standpoint. We will end this section with thoughts on potential innovative technologies that present a promising future in the context of COVID-19. A word of caution here – this study is not exhaustive, and we have included select cases/ examples that underscore our point about significant benefits that can be reaped in testing by the use of powerful technological interventions.

3.1   Current Experimentations

Biosensors   Researchers who had established the detection of bacteria and viruses in the air using biosensors have extrapolated that knowledge for the detection of the SARS-CoV-2 in the environment. These might be used in the future in crowded and most vulnerable places, like hospitals and laboratories, to detect the virus (Balfour 2020).

Nanotechnology   Nanoparticles and nanotechnology can help improve the diagnostic efficacy of current tests for detection and management. A more sensitive and less time-consuming approach has been suggested using a combination of functionalized nanoparticles and PCR-based assay (Gong et al. 2008). Nanoparticles can be made to attach to SARS-CoV-2. Then with the help of infrared light treatment, viral structure could be disrupted. This in turn will help to curb the development and survivability of the virus (Makichuk 2020). This has been done previously for microbes causing influenza and tuberculosis.

Point of Care Testing for Prevalence   Many countries, including India, have planned at implementing immunochromatographic test (ICT)-based kits for detection of IgG antibodies against SARS-CoV-2 in the blood sample. The results of an individual test can be obtained within 15 minutes. This will help to evaluate the prevalence of the infection in the population. Also, Enzyme-Linked Immunosorbent Assay (ELISA)-based serological diagnostic tests are being developed, which could be economical and done on a larger scale to detect a greater number of cases simultaneously (Ketchell 2020; Times Now 2020).

We would like to highlight efforts by a few countries (Taiwan, Israel, and South Korea) that have effectively used technology as a tool to help combat coronavirus.

3.1.1   Taiwan

Despite its proximity to China, Taiwan (population of ~24 million people), with 442 cases and 7 deaths as of May 31, 2020 (Taiwan News 2020), has so far been able to prevent the spread of COVID-19. Backed by its experience and learnings from past outbreaks, specifically SARS (2003) and H1N1 (2009), Taiwan was able to take rapid action during the initial stages of the spread. While there are numerous technological interventions, Taiwan relied on few of them, such as the Big Data – AI-powered proactive identification system and surveillance system deserve to mention. Taiwan integrated its National Health Insurance (NHIA) database, that contains health records of the majority of its population, with its Immigration and Customs database to identify suspected cases, which were then further investigated. It also implemented contact tracing by getting location records of infected individuals from their telecommunication providers. On surveillance system, Taiwan implemented Digital Fencing Tracking System that uses mobile phone signals to monitor people under quarantine. If the quarantine person moves away from his/her address or turns off the phone, local police and officials are alerted to conduct physical checks (Perelman 2020).

3.1.2 Israel

Nicknamed as "Startup Nation", Israel has become a budding ground for breakthrough technology interventions to fight the new public health crisis. Healthcare Services, an Israel-based healthcare organization, can identify high-risk individuals through the effective use of AI-based algorithm on its medical records, who are then tested promptly. Immediate medical intervention is provided to the positive cases, thereby minimizing the risk of further spread (Ho 2020). Multiple other efforts are in pipeline, which underscores Israel's reliance on technology as a critical part of its strategy to overcome COVID-19.

3.1.3 South Korea

After an initial rise in the number of infections, South Korea has managed well to flatten the SARS-CoV-2 infection curve without resorting to strict measures, such as a lockdown. It has achieved this transformation primarily by conducting tests extensively, effectively tracking positive cases, and adhering to social distancing norms. Through its Smart Management System (SMS), South Korea has nearly perfected the science of contact tracing (Ketchell 2020). It deployed innovative methods of contact tracing by tracking credit card usage of positive cases, smartphone locations of infected individuals, and their CCTV visuals. Through the use of technology, it also shared locations of the infected people and encouraged others to test themselves, if they were in similar localities. A one-nation strategy, which relied on efficient use of resources, acted as a shot in the arm in South Korea's fight (Fleming 2020) against the virus.

3.1.4 Sample Collection Kiosk

Sample collection kiosk was initiated in South Korea and is used in certain Indian states, like Kerala, North-Eastern states (Indian Express 2020). This helps in preserving the resources (example – personal protective equipment) and simultaneously decreases the exposure of healthcare professionals during sample collection.

3.1.5 Sniffer Dogs for Patient Detection

Countries, like the US and UK, are starting to train dogs to identify the smell of a COVID-19 sample (BBC 2020; PTI 2020). Dogs are being trained to identify samples, like urine and sputum, of COVID-19 positive cases. Dogs being highly sensitive to smell can detect volatile substances present in the samples. Such experiments have been useful earlier for the detection of certain cancers, malaria, etc. These trained dogs can then be employed to screen samples, and positively screened samples can then be confirmed by other tests. Claims are this can be helpful during the pre-symptomatic phase.

3.1.6   AI for COVID-19 Disease Management

Artificial intelligence (AI) makes it possible for machines to learn from experience, adjust to new inputs, and perform human-like tasks. Through AI, computers can be trained to accomplish specific tasks by processing large amounts of data. To illustrate the point and its application to COVID-19 testing, let us look at two specific examples:

1. Researchers at King's College, London, Massachusetts General Hospital, and health science company ZOE have developed an AI diagnostic app that can predict whether someone is likely to have COVID-19 based on their symptoms. Users of this app reported their symptoms daily. With the use of AI and data, the research team found a wide range of symptoms compared to cold and flu, specifically they found loss of taste and smell (anosmia) that is a stronger predictor of COVID-19 than fever (King's College 2020);
2. AI systems, trained via machine learning, may be able to identify patients that are critical and need intensive care immediately, easing up the pressure on the specialist to make that decision. ECART is the name of an AI system already in use in the US, which predicts the likelihood of a patient developing a potential cardiac condition. It uses data from patients' electronic medical records (such as vital signs, lab results, and demographic information) to provide real-time risk scores for patients (Strickland 2020). The application of AI opens up a wide array of possibilities, as far as testing and other areas of prevention and cure are concerned.

3.1.7   Robotics

Amidst the growing urgency to combat novel coronavirus, robots have been clearly showing their prowess across the globe, both indoors and outside. Their usage has been wide-ranging from dealing with highly infectious individuals within hospital premises to delivering groceries at people's doorsteps. UVD Robots, for example, are highly effective in surface cleaning, killing 99.99% of bacteria and viruses (Innovation and Blockchain 2020). They are being used in hospitals along with places, such as airports, hotels, offices, etc. (Blake 2020). Developed in Austin, Texas, another robot named Moxi is a hospital robot assistant that helps perform non-patient facing tasks, such as delivering lab samples, collecting items from a central supply, removing soiled linen bags, etc. (Diligent Robots 2020). A robot named Tommy, developed in Italy, helps monitor patient's vital statistics and relays them to hospital staff, thereby limiting direct contact of doctors and nurses with sensitive patients. Tommy robots also have a touch-screen face that allows patients to record messages and transmit them to the doctors. Many countries are, thus, using robots for works, like sanitization, food and medicine delivery, monitoring vital signs in patients, and even assisting in border controls (Dasgupta 2020). Robots are also being used to a contactless discussion between patients and doctors via video chat. These developments highlight the extensive potential that robotics present in man's resolve to overcome COVID-19. These robots can be built with sensors that can, for example, detect volatile compounds produced in positive samples and

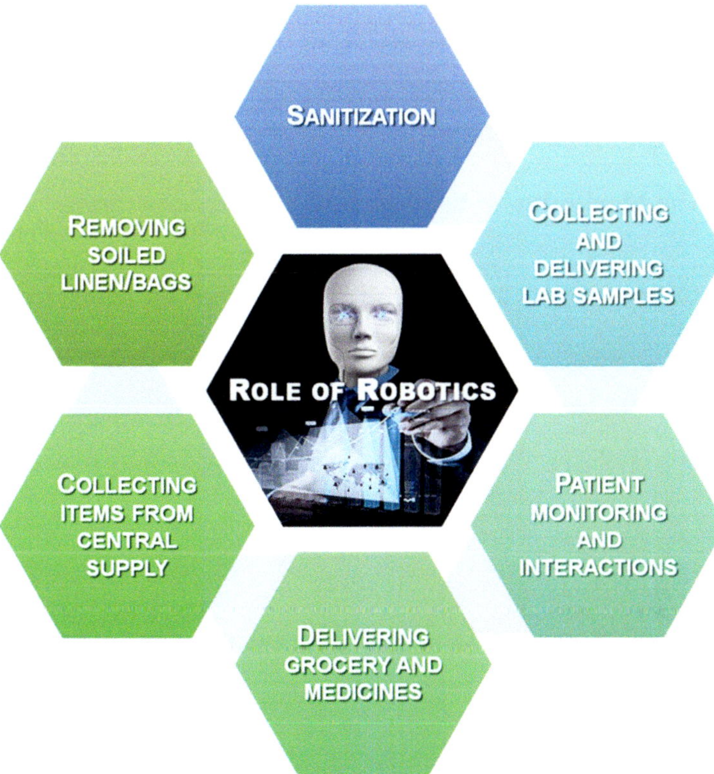

Fig. 3 Role of robotics during COVID-19

analyze them, preventing the exposure to the sniffer dogs. Robots can be designed to be used for sample collection, thus human exposure to the virus can be decreased (Fig. 3).

3.1.8 Drones

COVID-19 has pioneered innovative uses of *unmanned aerial vehicles*, (simply known as drones) across the globe. Our study on drones in the context of COVID 19 reveals that they are used for street surveillance and message broadcasting (Kuwait, UAE, Spain, etc.), spraying of disinfectants in affected areas (China, Indonesia, Chile, etc.), providing medical supplies to hospitals and labs (Wuhan), delivering groceries (China, US, Australia, etc.), and checking large-scale remote temperature (China). Also, drones are used by countries (Germany, the US, etc.) to survey areas, on which temporary hospitals are built (Sharma 2020). Besides the outlined applications, drones are believed to be more efficient and consistent; hence they are acting as a strong technological impetus in trying times like these. In India, they are being used to monitor quarantine centers and social distancing in crowded places apart from disinfecting COVID-specific areas of the hospital (Defense Aviation Post 2020).

3.2 Learning from Similar Illnesses in the Past

3.2.1 The Monitoring of Bat Population in the Local Areas

Because the virus was transmitted to humans from bats, periodic sampling and monitoring of the bat population in the nearby bat habitats can help predict any imminent outbreak of COVID-19 in the future.

3.2.2 Using Alternative Testing Methods

A machine named Gene Xpert, which was originally designed for detecting multidrug-resistant tuberculosis, was used in the detection of SARS cases. At present, also Gene Xpert and similar machines are being recommended by ICMR for detecting COVID-19 cases (WHO 2017). The research was done for developing tests that would give faster results than molecular diagnostic tests (James et al. 2018).

3.3 Innovations in Technology that Should Be Considered

3.3.1 Mobile Apps for Sending Sample Data to a Distant Location for Reporting

The research could focus on a device that could be connected to a smartphone app where the sample can be self-tested at home by the patient, and results can be interpreted by an expert via that app at a distant location.

3.3.2 Rapid Antigen Detection Kits for Viral Antigens in Urine

Kits are being designed by different companies to detect antigens in the sputum and throat swab samples. These kits detect the presence of viral proteins (antigens) expressed by the SARS-CoV-2 in the sample. Similar kits may be developed, where urine can be used as a sample. Dipstick kits can be a cheaper alternative to ICT kits for antigen detection.

3.3.3 Microarray for the Detection of Mutants

Microarray is a sophisticated molecular laboratory tool that can detect the expression of thousands of genes simultaneously. Because the SARS-CoV-2 is known to be highly mutating, this technique can be implemented in detecting the mutant genes and thus, specifically target these genes for detection and therapeutic intervention.

3.3.4 Use of Biometrics

As biometric devices are being used in face recognition sensors; they can be designed to have sensors to detect skin temperature and conjunctival scan-like things which can operate in a no-touch technique.

4 Conclusions

The COVID-19 pandemic has challenged the healthcare advancements of almost every country of the world. Both the developed and the developing countries have been struggling to fight with the virus. However, the characteristics, pathogenesis, treatment, and controlling measures are yet to come out. Till now, prompt detection and virus containment have been considered as the vital steps to deal with the virus. Asymptomatic spreaders are the key persons to identify to prevent contact and infection. However, no gold standard testing is available to do so, and the RT-PCR-based method also has limitations to be used in the same purpose. Several technological aspects, such as biosensors, nanotechnology, and point of care testing for prevalence, are being experimented to find out an ideal approach. Few suggested options, such as sniffer dogs for patient detection, AI, sample collection kiosk, robots, and drones, could also be tested. Robust replicative studies are necessary to focus the futuristic technological aspects for the early detection, prevention, and control of COVID-19, as some of the characteristics of the COVID-19 are still unknown.

Executive Summary
- COVID-19 pandemic has appeared as a major threat to the modern globalized civilization.
- Early diagnosis and virus containment are the vital steps to treat, control, as well as prevent the diseases.
- The identification of asymptomatic carriers is a fundamental step to prevent the spread of the virus.
- Different countries have been experimenting with different prospecting diagnostic tools where technology has been playing a crucial role.
- The RT-PCR-based method has limitations to be used in the primary prevention program.
- Biosensors, nanotechnology, and point of care testing for prevalence strategies are the ongoing experiments.
- Sniffer dogs for patient detection, artificial intelligence (AI), sample collection kiosks, robots, and drones could be potential options to be tested.
- Robust studies are warranted as some of the characteristics of the COVID-19 are still unknown.

References

Alizargar J, Etemadi SM, Aghamohammadi M, Hatefi S (2020) Saliva samples as an alternative for novel coronavirus (COVID-19) diagnosis. J Formos Med Assoc Taiwan Yi Zhi

Balfour H (2020) Developing a COVID-19 biosensor for faster diagnostics and environmental monitoring. https://www.europeanpharmaceuticalreview.com/news/117659/developing-a-covid-19-biosensor-for-faster-diagnostics-and-environmental-monitoring/

BBC (2020) Coronavirus: trial begins to see if dogs can "sniff out" virus. BBC News. https://www.bbc.com/news/uk-52686660

Bhusare BP, Zambare VP, Naik AA (2020) COVID-19: persistence, precautions, diagnosis, and challenges. J Pure Appl Microbiol. https://microbiologyjournal.org/covid-19-persistence-precautions-diagnosis-and-challenges/

Blake R (2020) In Coronavirus fight, robots report for disinfection duty. https://www.forbes.com/sites/richblake1/2020/04/17/in-covid-19-fight-robots-report-for-disinfection-duty/#4c0a671a2ada

Chan JF-W, Yuan S, Kok K-H, To KK-W, Chu H, Yang J, Xing F, Liu J, Yip CC-Y, Poon RW-S (2020) A familial cluster of pneumonia associated with the 2019 novel coronavirus indicating person-to-person transmission: a study of a family cluster. Lancet 395(10223):514–523

Dasgupta SS (2020) How robots are helping in fight against COVID-19. Geospatial World. https://www.geospatialworld.net/blogs/covid-19-how-robots-help-in-fight-against-covid-19/

Defense Aviation Post (2020) Here's how drones helping India in fight against coronavirus. https://www.defenceaviationpost.com/2020/04/heres-how-drones-helping-india-in-fight-against-coronavirus/

Diligent Robots (2020) Moxi helps hospitals and clinical staff. https://diligentrobots.com/moxi

Fleming S (2020) South Korea's Foreign Minister explains how the country contained COVID-19 [World Economic Forum]. https://www.weforum.org/agenda/2020/03/south-korea-covid-19-containment-testing/

Gong P, He X, Wang K, Tan W, Xie W, Wu P, Li H (2008) Combination of functionalized nanoparticles and polymerase chain reaction-based method for SARS-CoV gene detection. J nanosci nanotech 8(1):293–300

Ho D (2020) Israel proving to be med-tech hub during COVID-19 pandemic. BioWorld. https://www.bioworld.com/articles/434840-israel-proving-to-be-med-tech-hub-during-covid-19-pandemic

Indian Express (2020) Inspired by South Korea, Kerala launches walk-in Covid-19 sample collection kiosk. Indian Express. https://indianexpress.com/article/coronavirus/this-kerala-district-has-launched-an-innovative-model-for-faster-covid-19-sample-collection-6349837/

Innovation and Blockchain (2020) Danish disinfection robots save lives in the fight against the Corona virus. https://ec.europa.eu/digital-single-market/en/news/danish-disinfection-robots-save-lives-fight-against-corona-virus

James AS, Todd S, Pollak NM, Marsh GA, Macdonald J (2018) Ebolavirus diagnosis made simple, comparable and faster than molecular detection methods: preparing for the future. Virol J 15. https://doi.org/10.1186/s12985-018-0985-8

Ketchell M (2020) How South Korea flattened the coronavirus curve with technology. The Conversation. https://theconversation.com/how-south-korea-flattened-the-coronavirus-curve-with-technology-136202

King's College (2020) New AI diagnostic can predict COVID-19 without testing [Sciencedaily]. https://www.sciencedaily.com/releases/2020/05/200511112628.htm

Mahapatra S, Chandra P (2020) Clinically practiced and commercially viable nanobio engineered analytical methods for COVID-19 diagnosis. Biosens Bioelectron 165:112361

Makichuk, D (2020) Nanotechnology could aid in Covid-19 fight. Asia Times. https://www.asiatimes.com/2020/04/nanotechnology-could-aid-in-covid-19-fight

Padhi A, Kumar S, Gupta E, Saxena SK (2020) Laboratory Diagnosis of Novel Coronavirus Disease 2019 (COVID-19) Infection. In: Saxena S. (ed) Coronavirus Disease 2019 (COVID-19). Medical Virology: From Pathogenesis to Disease Control. Springer, Singapore. https://doi.org/10.1007/978-981-15-4814-7_9

Pascarella G, Strumia A, Piliego C, Bruno F, Del Buono R, Costa F, Scarlata S, Agrò FE (2020) COVID-19 diagnosis and management: a comprehensive review. J Intern Med 288:192–206

Perelman, M. (2020). Taiwan's vice president says 'possibility' that Covid-19 came from Chinese laboratory. The Interview. France 24. https://www.france24.com/en/asia-pacific/20200508-taiwan-s-vice-president-says-possibility-that-covid-19-came-from-chinese-laboratory [Accessed on 29 May 2020].

PTI (2020) Scientists believe trained dogs will start sniffing out coronavirus patients by July Read more at: https://economictimes.indiatimes.com/magazines/panache/scientists-believe-trained-dogs-will-start-sniffing-out-coronavirus-patients-by-july/articleshow/75533707.cms?utm_source=contentofinterest&utm_medium=text&utm_campaign=cppst.%20The%20Economic%20Times.%20https://economictimes.indiatimes.com/magazines/panache/scientists-believe-trained-dogs-will-start-sniffing-out-coronavirus-patients-by-july/articleshow/75533707.cms

Ravichandran K, Anbazhagan S, Agri H, Rupner RN, Rajendran VKO, Dhama K, Singh BR (2020) Global status of COVID-19 diagnosis: an overview. J Pure Appl Microbiol 14:879–892

Sabino-Silva R, Jardim ACG, Siqueira WL (2020) Coronavirus COVID-19 impacts to dentistry and potential salivary diagnosis. Clin Oral Investig 24:1–3

Sharma M (2020) How drones are being used to combat COVID-19. Geospatial World. https://www.geospatialworld.net/blogs/how-drones-are-being-used-to-combat-covid-19/

Shuja J, Alanazi E, Alasmary W, Alashaikh A (2020) COVID-19 datasets: a survey and future challenges. MedRxiv

Soraya GV, Ulhaq ZS (2020) Crucial laboratory parameters in COVID-19 diagnosis and prognosis: an updated meta-analysis. Available at SSRN 3576912

Strickland E (2020) AI can help hospitals triage COVID-19 patients. https://spectrum.ieee.org/the-human-os/artificial-intelligence/medical-ai/ai-can-help-hospitals-triage-covid19-patients

Taiwan News (2020) Sunday marks 49th consecutive day Taiwan goes without domestically transmitted COVID-19 cases. By George Liao. https://www.taiwannews.com.tw/en/news/3942497

Times Now (2020) IIT Delhi gets financial support from Microsoft India for two projects on COVID-19 detection research. https://www.timesnownews.com/education/article/iit-delhi-gets-financial-support-from-microsoft-india-for-two-projects-on-covid-19-detection-research/598365

Tozzi A, Amato G, Guarino A (2020) Rectal swabs for Covid-19 diagnosis. https://doi.org/10.13140/RG.2.2.16027.62246

WHO (2017) New technology allows for rapid diagnosis of Ebola in Democratic Republic of the Congo. World Health Organization. https://www.afro.who.int/news/new-technology-allows-rapid-diagnosis-ebola-democratic-republic-congo

Xu X, Yu C, Qu J, Zhang L, Jiang S, Huang D, Chen B, Zhang Z, Guan W, Ling Z (2020) Imaging and clinical features of patients with 2019 novel coronavirus SARS-CoV-2. Eur J Nucl Med Mol Imaging 47:1–6

Yang P, Wang X (2020) COVID-19: a new challenge for human beings. Cell Mol Immunol 17 (5):555–557

Younes N, Al-Sadeq DW, AL-Jighefee H, Younes S, Al-Jamal O, Daas HI, Yassine H, Nasrallah GK (2020) Challenges in laboratory diagnosis of the novel coronavirus SARS-CoV-2. Viruses 12(6):582

Zoabi Y, Shomron N (2020) COVID-19 diagnosis prediction by symptoms of tested individuals: a machine learning approach. MedRxiv

Zu ZY, Jiang MD, Xu PP, Chen W, Ni QQ, Lu GM, Zhang LJ (2020) Coronavirus disease 2019 (COVID-19): a perspective from China. Radiology 296:200490

Next-Generation Rapid Advanced Molecular Diagnostics of COVID-19 by CRISPR-Cas

Ashish Srivastava, Taruna Gupta, Swatantra Kumar, and Shailendra K. Saxena ®

Abstract

The world is looking towards the development of early diagnosis, treatment, and prevention of Coronavirus disease (COVID-19) caused by SARS-CoV-2 in order to restrict its rapid transmission and mortality among the human population. Globally, more than 5 million cases have been reported with 0.3 million deaths by the end of May 2020. Currently, the World Health Organization (WHO) adopted the screening and diagnosis of SARS-CoV-2 infection with quantitative RT-PCR (qRT-PCR)-based kits; however, the suitability of such kits is restricted due to the requirement of specialized instruments, well-trained personnel, and unavailability in resource-limited areas. The CRISPR-Cas system has recently emerged as a versatile tool for medical research for gene editing, epigenetic control, and disease diagnosis. The use of CRISPR-Cas-based detection of SARS-CoV-2 infection may results in the development of rapid, affordable, and multiplexed point-of-care diagnostic system with high specificity and sensitivity. In this article, we have covered the CRISPR-Cas-based efficient techniques developed for the diagnosis of the SARS-CoV-2 and their suitability for COVID-19 surveillance.

Keywords

CRISPR-associated proteins · Cas9 · Cas12 · Cas13 · Isothermal amplification · Point-of-care testing

A. Srivastava · T. Gupta
Amity Institute of Virology & Immunology, Amity University Uttar Pradesh, Noida, India

S. Kumar · S. K. Saxena (✉)
Centre for Advanced Research (CFAR), Faculty of Medicine, King George's Medical University (KGMU), Lucknow, India
e-mail: shailen@kgmcindia.edu

P. Chandra, S. Roy (eds.), *Diagnostic Strategies for COVID-19 and other Coronaviruses*, Medical Virology: From Pathogenesis to Disease Control, https://doi.org/10.1007/978-981-15-6006-4_9

Abbreviations

CARMEN	Combinatorial Arrayed Reactions for Multiplexed Evaluation of Nucleic acids
Cas	CRISPR-associated proteins
CASLFA	Clustered Regularly Interspaced Short Palindromic Repeats/Cas9-Mediated Lateral Flow Nucleic Acid Assay
COVID-19	Coronavirus disease
CRISPR	Clustered regularly interspaced short palindromic repeats
dCas9	Nuclease-deactivated Cas9
FELUDA	FnCas9 Editor-Linked Uniform Detection Assay
FnCas9	*Francisella novicida* Cas9
HEPN	Higher eukaryotes and prokaryotes nucleotide-binding domain
HOLMES	One-hour low-cost multipurpose highly efficient system
HUDSON	Heating unextracted diagnostic samples to obliterate nucleases
LAMP	Loop-mediated isothermal amplification
NASBACC	Nucleic acid sequence-based amplification-CRISPR cleavage
NHEJ	Non-homologous end joining
Nsp	Non-structural proteins
PAM	Protospacer adjacent motif
PC	Paired dCas9
RCA	Rolling circle amplification
RCH	RCA-CRISPR-split-HRP
RPA	Recombinase polymerase amplification
RT-qPCR	Quantitative polymerase chain reaction
SARS-CoV-2	Severe Acute Respiratory Syndrome Coronavirus 2
sgRNA	Single guide RNA
SHERLOCK	Specific high sensitivity enzymatic reporter unlocking
SpCas9	*Streptococcus pyogenes* Cas9

1 Introduction

Coronavirus disease (COVID-2019) is caused by a novel coronavirus, (SARS-CoV-2) – a large known enveloped RNA virus that causes respiratory illnesses of varying severity from the common cold to fatal pneumonia. SARS-CoV-2 is a newly spilled coronavirus into humans which is closely related to SARS-CoV (Kumar et al. 2020a, b). The genomic RNA of coronaviruses is around 30 kb comprising of 11 open reading frames (27 proteins) that encodes for 4 major viral structural proteins envelope, spike, membrane, and nucleocapsid (N) (Khailany et al. 2020) which are involved in SARS-CoV-2 attachment and internalization, while other major ORF1a and ORF1b encode 16 non-structural proteins (NSP1–NSP16) that helps in viral replication and transcription (Helmy et al. 2020) (Kumar et al. 2020a,b). It has been reported to infect humans, bats, mice, and other animal's respiratory, nervous, and gastrointestinal systems. The incubation period is ranging from 2–14 days for

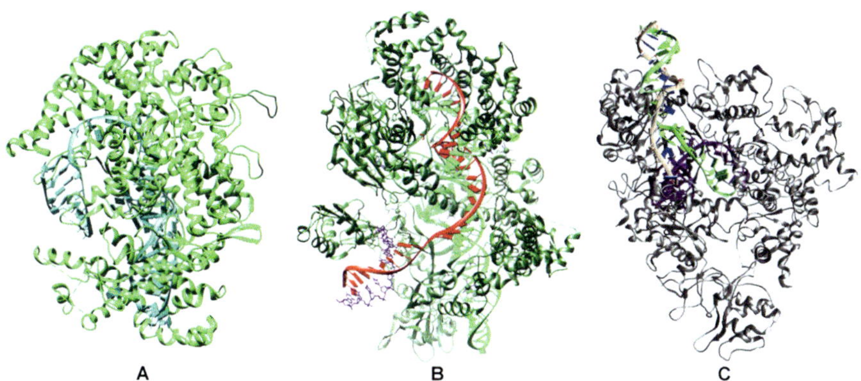

Fig. 1 Various CRISPR-associated proteins (Cas), such as Cas13a (A), F9Cas9 (B), and Cas12a (C) to be used in diagnosis via nucleic acid testing

COVID-19 symptoms to arise. The symptoms of COVID-19 include respiratory illness with fever, cough, and shortness of breath which are similar to a common flu (Cascella et al. 2020). However, severe cases of COVID-19 represent varied symptoms, including acute respiratory distress syndrome, chest pain, organ failure, and death. Patients with mild symptoms or asymptomatic are acting as carriers who are unknowingly transmitting the virus. In order to prevent the SARS-CoV-2 transmission, rapid diagnostic kits and disease monitoring and surveillance program in large population needs to be conducted to minimize the number of new cases.

The current diagnosis of COVID-19 relies on the quantitative polymerase chain reaction (RT-qPCR). However, the suitability of such technique is restricted due to the requirement of specialized instruments, well-trained personnel and unavailability in resource-limited areas. Thus, there is a need to develop effective diagnostic tools with high accuracy and sensitivity and easy to use for detection of SARS-CoV-2 infection. Clustered regularly interspaced short palindromic repeats (CRISPR)-Cas system is rapidly becoming prevalent for its powerful potential in therapeutics but also getting unique identity in the field of molecular diagnostics. Understanding of diverse CRISPR-Cas systems has expanded its application in genome-wide screening, multiplex genome editing, transcriptional regulation, gene therapy, and antiviral therapy. With the recent discovery of various CRISPR-associated proteins (Cas), some of them, such as Cas13a (PDB:5XWY), F9Cas9 (PDB:5B2O), and Cas12a (PDB:6GTC), may be used to for the development of reliable, highly sensitive, cost-effective diagnostics through nucleic acid screening (Fig. 1).

2   Screening and Diagnostics for COVID-19 Patients

The most quick screening for COVID-19 suspects is mass thermal scanning-based temperature detection, which detects fever in the suspected individuals. However, asymptomatic individuals or people who have taken antipyretic medication may

Table 1 CRISP-Cas-based biosensing system

Classification	System name	Effector	Signal amplification
Cas9	RCH	Sp-dCas9	RCA
	NASBACC	SpCas9	NASBA
	PC reporter	Sp-dCas9	PCR
	CAS-EXPAR	SpCas9	EXPAR
Cas12	HOLMES	LbCas12a	PCR; RT-PCR
	DETECTR	LbCas12a	RPA
	HOLMESv2	AacCas12b	LAMP; RT-LAMP; Asymmetric PCR
Cas13	SHERLOCK	LwCas13a	RPA
	HUDSON +SHERLOCK	LwCas13a	RPA
	SHERLOCKv2	CcaCas13b, PsmCas13b, LwaCas13a	RPA

RCH RCA-CRISPR-split-HRP, *NASBA* Nucleic acid sequence-based amplification, *NASBACC* Nucleic acid sequence-based amplification-CRISPR cleavage, *PC* Paired dCas9, *RPA* Recombinase polymerase amplification, *sgRNA* Single guide RNA, *SHERLOCK* Specific high sensitivity enzymatic reporter unlocking, *LAMP* Loop-mediated isothermal amplification, *HOLMES* One-hour low-cost multipurpose highly efficient system, *dCas9* Nuclease-deactivated Cas9, *RCA* Rolling circle amplification, *HUDSON* Heating unextracted diagnostic samples to obliterate nucleases, *DETECTR* DNA Endonuclease-Targeted CRISPR Trans Reporter, *EXPAR* Exponential amplification reaction

escape from this primary screening and therefore remains as the carriers of virus transmission. The current molecular diagnosis of COVID-19 relies on the RT-qPCR-based amplification of viral RNA gene from nasopharyngeal or oropharyngeal specimens collected from the suspected individuals (Corman et al. 2020). This test uses three primer and probe sets to detect three regions as nucleocapsid (N) gene and one primer and probe set to detect human RNase P (RP) in clinical samples. The isothermal alternatives of PCR, such as RPA (Recombinase Polymerase Amplification) and LAMP (Loop-Mediated Isothermal Amplification)-based assays, are also being standardized for diagnosis of SARS-CoV-2 (Park et al. 2020; Yu et al. 2020; Yinhua Zhang et al. 2020). These sensitive molecular diagnostic technologies are rapid and portable and can be multiplexed for several pathogens (Rostron et al. 2019; Miao et al. 2019; Notomi et al. 2000). The detailed analysis of these molecular diagnostic mechanisms, low sensitivity, and low throughput were observed (Huang et al. 2020). Interestingly, CRISPR-Cas-based nucleic acid detection kit has been developed for rapid, low-cost diagnosis of Zika virus from the plasma (Pardee et al. 2016). Apart from the remarkable gene editing application, CRISPR-Cas system may be used in biosensing applications (Table 1). CRISPR-Cas-based biosensing system can be classified into three major groups on the basis of Cas effectors (Li et al. 2019a,b).

3   CRISPR-Cas Technology for Diagnostics

CRISPR (clustered regularly interspaced short palindromic repeats) is a family of DNA sequences found within the genome of prokaryotic organisms. Cas (CRISPR-associated protein), the most commonly used endonuclease, utilizes a guide RNA to bind to a complementary DNA sequence, which is subsequently cleaved through Cas endonuclease activity. Cas9 recognizes specific protospacer adjacent motif (PAM), a 2–6 bases on the target sequences upstream to guide RNA target. The most commonly studied endonuclease belongs to type II CRISPR system, i.e., Cas9 that targets DNA phage's while the type V CRISPR system, i.e., Cas12 targets both ssDNA and dsDNA, and type III and type VI groups exhibit RNA targeting activity which includes Cas13. These activities can be multiplexed with each other for development of more reliable diagnostic kits with higher specificity.

Cas9 can recruit guide RNA (sgRNA) that facilitates its specific binding to target DNA and can create a blunt dsDNA break or nick, which repaired by cellular machinery either through non-homologous end joining (NHEJ) or homology directed repair (HDR) when coupled with a single-stranded or double-stranded donor DNA to create site-specific edits. Cas9s are preferred for its high genome editing efficiency; however, off-target cleavage at unintended genome sites can be a disadvantage for in vivo applications. The recruitment of the PAM site on the target DNA is necessary for Cas9 binding. Most commonly used *Streptococcus pyogenes* Cas9 (SpCas9) recognizes 5' NGA 3', and *Francisella novicida* Cas9 (FnCas9) recognizes 5' NGG 3' as the PAM sequence on the target site (Wiedenheft et al. 2012; Hirano et al. 2016). Another form of Cas9 protein, that is the dCas9 (nuclease deficient Cas9 or dead Cas9), is being applied to suppress gene expression when applied to the transcription binding site of the desired part of a gene. The specific binding feature of dCas9 with sgRNA is generally being utilized for diagnostics purposes (Zhou et al. 2018a).

Another class of Cas protein (Class 2 type V) is Cas12 (Cpf1), which is a proficient enzyme that creates staggered cuts in dsDNA (Chen et al. 2018). Cas12 develops its own crRNAs that leads to increased multiplexing ability. Cas12 has brought a platform for epigenome editing. Cas12a can slice ssDNA once activated by a target DNA molecule identical to its spacer sequence (Chen et al. 2018; Swarts et al. 2017). Unlike Cas9, Cas12a recognizes a T-rich PAM instead of a G-rich PAM and generate dsDNA breaks with staggered 5' ends (Stella et al. 2018). CRISPR-Cas12a possesses the cis-trans cleavage activity of on ssDNA which is being used for diagnosis of various pathogens (Chen et al. 2018) (Fig. 2).

Cas13 (C2c2) are unique CRISPR-associated effectors which have specific recognition and cleavage activity for complementary RNA and known for trans cleavage or collateral cleavage of nearby RNA (Abudayyeh et al. 2016). This feature is widely being utilized for the diagnosis of viruses and other miRNAs (Kellner et al. 2020). The class of protein contains Cas13a, Cas13b, Cas13c, and Cas13d, which can be engineered to cleave target RNA strands (Smargon et al. 2020). These Cas13 orthologues have been recognized and utilized for specific RNA modification in both prokaryotic and eukaryotic systems, such as RNA-dependent editing and detection

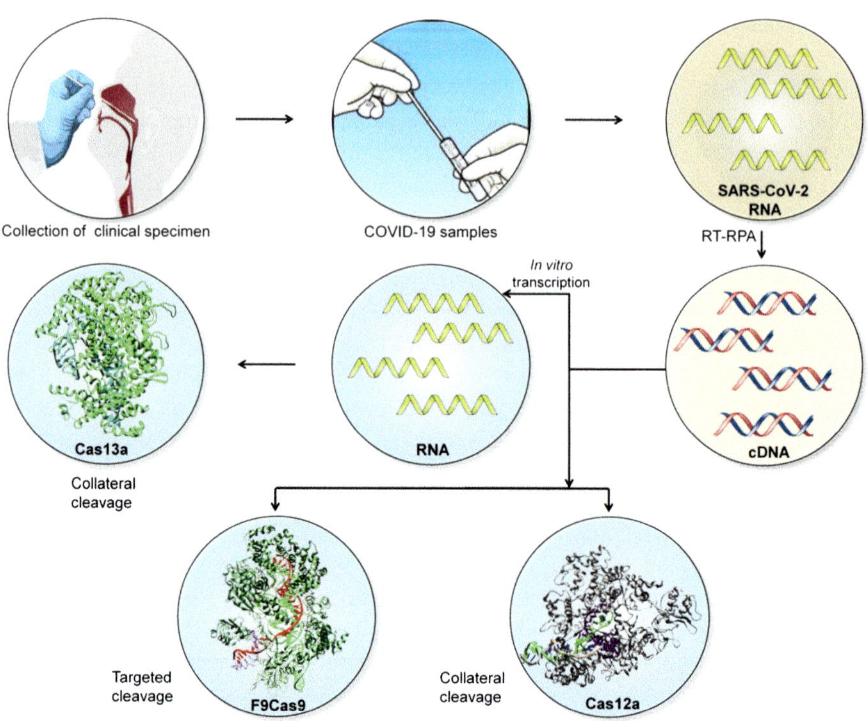

Fig. 2 Nasopharyngeal or oropharyngeal specimens collected from the suspected individuals are subject to reverse transcription recombinase polymerase amplification (RT-RPA) of SARS-CoV-2 RNA to cDNA. The cDNA can be converted to RNA via in vitro transcription using T7 RNA polymerase and subjected to collateral cleavage via Cas13a. Similarly, the cDNA can be directly subjected to targeted or collateral cleavage via F9Cas9 and Cas12a, respectively, in presence of reporter molecules

mechanisms (Liu et al. 2017). Cas13 with crRNA, once bind to it target ssRNA and get activated, it releases a nonspecific RNase activity and cleaves nearby RNAs regardless of their sequence. This property is being utilized in vitro for diagnostic development by coupling it with some non-complementary probe RNA. They contain the bilobed domain architecture of a recognition lobe (REC lobe) and a nuclease lobe (NUC lobe). The higher eukaryotes and prokaryotes nucleotide-binding (HEPN) domain along with the helical domain are involved in recognition of crRNA-target duplex in Cas13d, Cas13a while an additional RRI-1 domain is associated for recognition by Cas13b (Gootenberg et al. 2017).

In the scenario of COVID-19 outbreak, reliable, quick, and highly sensitive molecular diagnostics are very essential. CRISPR-based diagnostic tools may play an important role in development of such diagnostics. These tools were combined with various platforms of amplification, like RPA and LAMP, and also multiplexed with various devices for detection, such as plate readers, fluorometer, and lateral flow devices, to make them accessible and can be applied at the point of operation.

Recent report of rolling circle amplification (RCA)-assisted CRISPR-Cas9 cleavage (RACE) also suggest the power of CRISPR-Cas9 for multiplexed and specific detection of nucleic acids in point-of-care diagnostics (Wang et al. 2020a). CRISPR-Cas systems have displayed a remarkable potential platform for development of next-generation biosensing applications for highly sensitive nucleic acid diagnosis owing to the collateral cleavage activity of Cas effector proteins (Cas12 and Cas13) in very less time, providing attomolar sensitivity, single nucleotide specificity, and easy to use detection method. Here we are highlighting major CRISPR-Cas-based assays developed for diagnostic purposes with special emphasis on COVID-19 (Table 2).

4   CRISPR-Associated Proteins (Cas) for COVID 19 Diagnostics Development

4.1   Cas9-Based Diagnostics

The first CRISPR-Cas9-based diagnostics were developed coupled with an isothermal amplification technique, nucleic acid sequence-based amplification (NASBA) to distinguish in vitro the closely related Zika virus strains (Pardee et al. 2016). In this diagnostic method, RNA is extracted and amplified via NASBA and rehydrated the freeze-dried paper sensors, and the detection of the target RNA was indicated by a color change in the paper disc from yellow to purple. Later, Zhou et al. (2018b) developed a CRISPR-Cas9-triggered nicking endonuclease-mediated strand displacement amplification method (namely CRISDA) for ultrasensitive, point-of-care diagnostics and field analyses. Wang et al. (2020b) have come up with a unique diagnostic assay named CRISPR-Cas9-Mediated Lateral Flow Nucleic Acid Assay (CASLFA) for diagnosis of African swine fever virus (ASFV) by lateral flow nucleic acid diagnosis kit and in this proposal, and we have adopted the idea with this method (Fig. 3). Recently, a Cas9-based rapid diagnostic kit has been developed for COVID-19 where a highly accurate enzymatic readout FnCas9 Editor Linked Uniform Detection Assay (FELUDA) was used for detection of nucleotide sequences. This approach is based on the specific binding of sgRNA and dCas9 with target dsDNA; therefore, it may prove highly sensitive technique for diagnosis of nucleic acids. RT-RPA amplification and dFaCas9 have been combined for the development of lateral flow device for SARS-CoV-2 RNA detection.

4.2   Cas12-Based Diagnostics

Cas12 proteins are around 1300 amino acid long which is being used for CRISPR-based diagnosis. CRISPR-Cas12-based diagnostic kits were also made in lateral flow devices as well as fluorometer based diagnosis (Yan et al. 2019). The Cas12a-based system utilizes an ssDNA probes in place of RNA probes, therefore, is more suitable for on-site viral detection. HOLMES (one-hour low-cost multipurpose highly

Table 2 Various platforms for viral RNA diagnosis, their utility and sensitivity

Platform	Sensitivity	Time (approx.)	Cost effective	Extensive training	Single nucleotide specificity	References
Cas9 (CASLFA, FELUDA)	High	60 MINS	YES	NO	YES	Pardee et al. (2016) and Azhar et al. (2020)
Cas12 (DETECTR)	Atto-molar	40 MINS	YES	NO	YES	Broughton et al. (2020) and Bai et al. (2019)
Cas13 (SHERLOCK)	Atto-molar	60 MINS	YES	NO	YES	Rauch et al. (2020) and Kellner et al. (2020)
qRT-PCR	Low	120 MINS	NO	YES	NO	Corman et al. (2020)
RT- LAMP	Low	120 MINS	NO	YES	NO	Notomi et al. (2000)

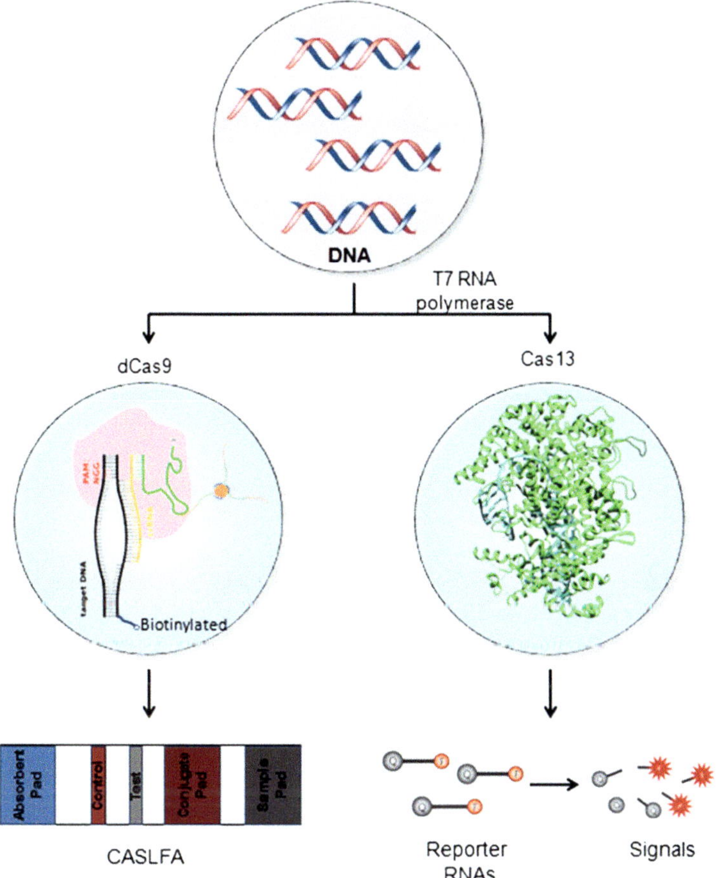

Fig. 3 Amplified DNA may be subjected to dCas9 binding through CASLFA or transcribed to RNA and targeted with Cas13 in SHERLOCK

efficient system) and DETECTR (DNA Endonuclease-Targeted CRISPR Trans Reporter) are two major CRISPR-Cas12-based diagnostic system which has been applied worldwide (Li et al. 2018; 19) coupled with isothermal amplification. A recent report of Broughton et al. (2020) demonstrated CRISPR-Cas12-based detection of SARS-CoV-2 utilizing DETECTR assay and claimed as a visual and faster alternative to the SARS-CoV-2 real-time RT-PCR assay. This assay offers a colorimetric and rapid alternative to for SARS-CoV-2 real-time RT-PCR assay, with 95% positive predictive agreement and 100% negative predictive agreement. Sensitivity of CRISPR-Cas12-mediated diagnosis demonstrates better specificity than other Cas-mediated diagnosis in a recent report (Huang et al. 2020). Capacity of Cas12 proteins to recognize and cleave both dsDNA and ssDNA makes it unique to diagnose a wide range of pathogens in crude samples.

4.3   Cas13-Based Diagnostics

CRISPR-Cas13 has played a vital role in viral diagnostic using platform specific high-sensitivity enzymatic reporter unlocking (SHERLOCK). The target RNA either detected directly in one step or coupled with RT-RPA isothermal amplification in two steps. Cas13 binds with the target RNA sequences, cleaves surrounding RNA transcripts, including the RNA reporter, due to its nonspecific cleavage property which results in the emission of fluorescent signal that recorded by the detector (Kellner et al. 2020). For diagnosis of SARS-CoV-2, CRISPR-Cas13-based diagnostics have explored much for nucleic acid-based diagnosis. Researchers have developed a protocol of the application of SHERLOCK for COVID-19 diagnosis (Fig. 1) (Zhang et al. (2020). A novel method based on Cas13-based effector that is rugged, equitable, scalable testing (CREST) has been developed. This method is required low-cost instrumentation without losing detection sensitivity (Rauch et al. 2020). The ultimate detection in these methods is based on collateral cleavage of reporter RNA. A highly multiplexed method for pathogen detection was developed by combination of combinatorial arrayed reactions for multiplexed evaluation of nucleic acids (CARMEN) with Cas13 and demonstrated to detect 169 human viruses, including SARS-CoV-2 (Ackerman et al. 2020). Cas13-based diagnostics seems promising for future nucleic acid-based diagnostics due to its role in multiplexing and throughput diagnostic techniques development.

5   Conclusions

COVID-19 is a pandemic with more than 5 million total cases and more than 0.3 million total deaths all over the world. Among these cases, only the United States of America (USA) and European countries contributed for more than half of the cases and contain a number of asymptomatic individuals acting as silent spreaders. Therefore, this is an alarming situation and required to diagnose a high number of populations who are suspected to be a spreader of COVID-19. To estimate the actual active cases worldwide, there is a need of inexpensive, rapid, and reliable diagnostic techniques for surveillance of the disease in the mass population, and RT-PCR-based diagnostic technique may prove inefficient for this purpose due to high infrastructural arrangements costs. Based on the study, the CRISPR-Cas-based nucleic acid detection technique may be proving significant for the current situation. Multiplexing of CRISPR-Cas-based diagnostics plays reasonable role to prove it a next-generation system for disease surveillance. Evidence for multiplexed diagnosis of more than 160 viruses with low infrastructural requirements shows its usability as point-of-care testing system.

6   Future Perspectives

Advancement in the current researches on CRISPR-Cas technique shows its potential to become the next-generation diagnostic tool for early, rapid, and reliable nucleic acid-based diagnostics. Trials for the CRISPR-Cas-based diagnosis already being performed, and these diagnostics are found potential for SARS-CoV-2 diagnosis. Recently, the Emergency Use Authorization (EUA), Food and Drug Administration (FDA), and the USA has already approved its first CRISPR-Cas13 (SHERLOCK)-based diagnostic kit for early detection of SARS-CoV-2 infection, which shows the potential of this tool to become the next-generation diagnostic kits.

Acknowledgment and Disclosures Authors are thankful to the Department of Science and Technology (DST-INSPIRE-Faculty Scheme, IFA-17-LSPA-100) for grant to A. Srivastava, Indian Council of Medical Research (ICMR)-SRF (2020–7629) to T. Gupta. The authors are also grateful to the Vice Chancellor, King George's Medical University (KGMU), Lucknow, India, for encouraging this work. The authors have no other relevant affiliations or financial involvement with any organization or entity with a financial interest in or financial conflict with the subject matter or materials discussed in the manuscript apart from those disclosed.

References

Abudayyeh OO, Gootenberg JS, Konermann S et al (2016) C2c2 is a single-component programmable RNA-guided RNA-targeting CRISPR effector. Science 353:aaf5573

Ackerman CM, Myhrvold C, Thakku SG, et al. (2020) Massively multiplexed nucleic acid detection with Cas13 [published online ahead of print, 2020 Apr 29]. Nature https://doi.org/10.1038/s41586-020-2279-8

Azhar A, Phutela R, Ansari AH et al (2020) Rapid, field-deployable nucleobase detection and identification using FnCas9. bioRxiv. https://doi.org/10.1101/2020.04.07.028167

Bai J, Lin H, Li H, Zhou Y, Liu J, Zhong G (2019) Cas12a-based on-site and rapid nucleic acid detection of African swine fever. Front Microbiol 10(10):2830

Broughton JP, Deng X, Yu G, et al. (2020) CRISPR-Cas12-based detection of SARS-CoV-2 [published online ahead of print, 2020 Apr 16]. Nat Biotechnol https://doi.org/10.1038/s41587-020-0513-4

Cascella M, Rajnik M, Cuomo A, et al. (2020) Features, evaluation and treatment Coronavirus (COVID-19). In: StatPearls [Internet]. StatPearls Publishing, Treasure Island (FL)

Chen JS, Ma E, Harrington LD, Da Costa M, Tian X, Palefsky JM, Doudna JA (2018) CRISPR Cas12a target binding unleashes indiscriminate single-stranded DNase activity. Science 360 (6387):436–439

Corman VM, Landt O, Kaiser M et al (2020) Detection of 2019 novel coronavirus (2019-nCoV) by real-time RT-PCR. Euro Surveill 25(3):2000045

Gootenberg JS, Abudayyeh OO, Lee JW et al (2017) Nucleic acid detection with CRISPR-Cas13a/C2c2. Science (New York, NY) 356(6336):438–442

Helmy YA, Fawzy M, Elaswad A, Sobieh A, Kenney SP, Shehata AA (2020) The COVID-19 pandemic: a comprehensive review of taxonomy, genetics, epidemiology, diagnosis, treatment, and control. J Clin Med 9(4):1225

Hirano S, Nishimasu H, Ishitani R, Nureki O (2016) Structural basis for the altered PAM specificities of engineered CRISPR-Cas9. Mol Cell 61(6):886–894

Huang Z, Tian D, Liu Y et al (2020) Ultra-sensitive and high-throughput CRISPR-powered COVID-19 diagnosis. Biosens Bioelectron 23:112316

Kellner MJ, Koob JG, Gootenberg JS, Abudayyeh OO, Zhang F (2020) SHERLOCK: nucleic acid detection with CRISPR nucleases. Nat Protoc 15(3):1311

Khailany RA, Safdar M, Ozaslan M (2020) Genomic characterization of a novel SARS-CoV-2 [published online ahead of print, 2020 Apr 16]. Gene Rep 19:100682

Kumar S, Maurya VK, Prasad AK, Bhatt MLB, Saxena SK (2020a) Structural, glycosylation and antigenic variation between 2019 novel coronavirus (2019-nCoV) and SARS coronavirus (SARS-CoV). Virus 31:13–21

Kumar S, Nyodu R, Maurya VK, Saxena SK (2020b) Morphology, genome organization, replication, and pathogenesis of Severe Acute Respiratory Syndrome Coronavirus 2 (SARS-CoV-2). In Coronavirus disease 2019 (COVID-19). Springer, Singapore, pp 23–31

Li SY, Cheng QX, Liu JK, Nie XQ, Zhao GP, Wang J (2018) CRISPR-Cas12a has both cis- and trans-cleavage activities on single-stranded DNA. Cell Res 28:491–493

Li L, Li S, Wu N, Wu J, Wang G, Zhao G, Wang J (2019a) HOLMESv2: a CRISPR-Cas12b-assisted platform for nucleic acid detection and DNA methylation quantitation. ACS Synth Biol 8(10):2228–2237

Li Y, Li S, Wang J, Liu G (2019b) CRISPR/Cas systems towards next-generation biosensing. Trends Biotechnol 37:730–743

Liu L, Chen P, Wang M, Li X, Wang J, Yin M et al (2017) C2c1-sgRNA complex structure reveals RNA-guided DNA cleavage mechanism. Mol Cell 65:310–322

Miao F, Zhang J, Li N, Chen T, Wang L, Zhang F (2019) Rapid and sensitive recombinase polymerase amplification combined with lateral flow strip for detecting african swine fever virus. Front Microbiol 15:1004

Notomi T, Okayama H, Masubuchi H, Yonekawa T, Watanabe K, Amino N, Hase T (2000) Loop-mediated isothermal amplification of DNA. Nucleic Acids Res 28:63

Pardee K, Green A, Takahashi MK, Braff D et al (2016) Rapid, low-cost detection of zika virus using programmable biomolecular components. Cell 165:1255–1266

Park GS, Ku K, Baek SH, Kim SJ, Kim SI, Kim BT, Maeng JS (2020) Development of reverse transcription loop-mediated isothermal amplification assays targeting SARS-CoV-2. J Mol Diagn S1525–1578:30090–30098

PDB ID: 5B2O, Hirano H, Gootenberg JS, Horii T et al (2016) Structure and engineering of *Francisella novicida* Cas9. Cell 164:950–961

PDB: 5XWY, Liu L, Li X, Ma J et al (2017) The molecular architecture for RNA-guided RNA cleavage by Cas13a. Cell 170:714–726.e10

PDB: 6GTC, Stella S, Mesa P, Thomsen J et al (2018) Conformational activation promotes CRISPR-Cas12a catalysis and resetting of the endonuclease activity. Cell 175:1856–1871.e21

Rauch JN, Valois E, Solley SC et al (2020) A scalable, easy-to-deploy, protocol for cas13-based detection of SARS-CoV-2 genetic material. bioRxiv. https://doi.org/10.1101/2020.04.20.052159

Rostron P, Pennance T, Bakar F et al (2019) Development of a recombinase polymerase amplification (RPA) fluorescence assay for the detection of Schistosoma haematobium. Parasit Vectors 12:514

Smargon AA, Shi YJ, Yeo GW (2020) RNA-targeting CRISPR systems from metagenomic discovery to transcriptomic engineering. Nat Cell Biol 22:143–150

Swarts DC, van der Oost J, Jinek M (2017) Structural basis for guide RNA processing and seed-dependent DNA targeting by CRISPR-Cas12a. Mol Cell 66:221–233e4

Stella S, Mesa P, Thomsen J, Paul B, Alcón P, Jensen SB, Saligram B, Moses ME, Hatzakis NS, Montoya G (2018) Conformational activation promotes CRISPR-Cas12a catalysis and resetting of the endonuclease activity. Cell 175(7):1856–1871.e21

Wang X, Xiong E, Tian T et al (2020a) Clustered regularly interspaced short palindromic repeats/Cas9-mediated lateral flow nucleic acid assay. ACS Nano 4:2497–2508

Wang R, Zhao X, Chen X et al (2020b) Rolling circular amplification (RCA)-assisted CRISPR/Cas9 cleavage (RACE) for highly specific detection of multiple extracellular vesicle MicroRNAs. Anal Chem 92:2176–2185

Wiedenheft B, Sternberg SH, Doudna JA (2012) RNA-guided genetic silencing systems in bacteria and archaea. Nature 482:331–338

Yan F, Wang W, Zhang J (2019) CRISPR-Cas12 and Cas13: the lesser known siblings of CRISPR-Cas9. Cell Biol Toxicol 35:489–492

Yinhua Zhang NO, Xiong J, Sun L et al (2020) Rapid molecular detection of sars-cov-2 (covid-19) virus RNA using colorimetric LAMP. medRxiv. https://doi.org/10.1101/2020.02.26.20028373

Yu L, Wu S, Hao X et al (2020) Rapid detection of COVID-19 coronavirus using a reverse transcriptional loop-mediated isothermal amplification (RT-LAMP) diagnostic platform [published online ahead of print, 2020 Apr 21]. Clin Chem:hvaa102

Zhang F, Abudayyeh OO, Gootenberg JS. (2020) A protocol for detection of COVID-19 using CRISPR diagnostics. A protocol for detection of COVID-19 using CRISPR diagnostics. p 8

Zhou L, Peng R, Zhang R, Li J (2018a) The applications of CRISPR/Cas system in molecular detection. J Cell Mol Med 22:5807–5815

Zhou W, Hu L, Ying L et al (2018b) A CRISPR–Cas9-triggered strand displacement amplification method for ultrasensitive DNA detection. Nat Commun 9:5012

Negative COVID-19 Test: What Next?

Sarvodaya Tripathy, S. M. Yasir Arafat, Sujita Kumar Kar, and Shailendra K. Saxena ⓘ

Abstract

The COVID-19 pandemic is one of the most devastating tragedies of this century. Over the past few months, it affected almost all the countries in the world. There has been a change in the recommendations for testing the suspected COVID-19 patients and contacts from time to time, globally. The status of the test (positive or negative) has significant health implications. In this chapter, we discuss the implications of a negative result in COVID-19 test, its diagnostic, clinical, and psycho-social implications.

Keywords

COVID-19 test · Negative test · Diagnosis · SARS-CoV-2

Authors Sarvodaya Tripathy, Sujita Kumar Kar, Shailendra K Saxena have equally contributed to this chapter as first author.

S. Tripathy (✉)
Department of Microbiology, Maharaja Krishna Chandra Gajapati Medical College & Hospital, Brahmapur, Ganjam, Odisha, India

S. M. Y. Arafat
Department of Psychiatry, Enam Medical College and Hospital, Dhaka, Bangladesh

S. K. Kar
Department of Psychiatry, King George's Medical University (KGMU), Lucknow, India

S. K. Saxena
Centre for Advanced Research (CFAR), Faculty of Medicine, King George's Medical University (KGMU), Lucknow, India

P. Chandra, S. Roy (eds.), *Diagnostic Strategies for COVID-19 and other Coronaviruses*, Medical Virology: From Pathogenesis to Disease Control, https://doi.org/10.1007/978-981-15-6006-4_10

1   Introduction

In the latest situation report on COVID-19 by the World Health Organization (WHO), more than 3 million confirmed cases and 0.26 million deaths have been reported globally (World Health Organization 2020a). COVID-19 first appeared in Hubei Province, China in December 2019. A cluster of patients was admitted with fever and respiratory symptoms showing lung opacities in CT-scan and were initially diagnosed to have pneumonia. A negative multiplex real-time polymerase chain reaction (RT-PCR) of common respiratory pathogen panels further suggested pneumonia caused by an unknown etiological agent (Udugama et al. 2020; Mahapatra and Chandra 2020). Analysis of bronchoalveolar lavage (BAL) fluid revealed a pathogen with a similarity of the genetic sequence with earlier detected coronaviruses, severe acute respiratory syndrome virus (SARS-CoV) and Middle East respiratory syndrome virus (MERS-CoV). It, however, showed maximum genetic similarity with bat coronavirus RaTG13 (Zhou et al. 2020). First, the virus was detected in the bronchoalveolar lavage (BAL) fluid taken from patients in Wuhan, China. Then the virus was cultured in human respiratory epithelial cell culture, and the supernatant was analyzed (after making it non-infective) using transmission electron microscopy.

Currently, the laboratory diagnosis of coronavirus disease 2019 (COVID-19) is mostly done using a nucleic acid amplification test (NAAT). Real-time reverse-transcription polymerase chain reaction (rRT-PCR) is a recommended test (World Health Organization 2020b). It involves a lot of crucial steps starting from the correct collection of samples to the final interpretation of the test results. Both nasopharyngeal and oropharyngeal swab (upper respiratory tract samples) are the recommended samples by the WHO (Government of India 2020; World Health Organization 2020b).

The properly collected swab has to be placed in a viral transport medium, packed properly, and transported in the cold chain. The sample is processed in laboratories with biosafety level 2 or equivalent. The viral genetic material, i.e., RNA, is extracted. Specific segments/genes of the SARS-CoV-2 from the extracted RNA are then amplified and detected by the rRT-PCR. Any fault in even one of these steps could lead to a false-negative result (World Health Organization 2020b).

2   Meaning of Negative COVID-19 Test (True Negative Versus False Negative)

A negative result may be a true negative or false negative. True negative means the patient is not harboring the virus, and the rRT-PCR is simultaneously showing an absence of viral genes. A false-negative result means that the patient is having COVID-19/SARS-CoV-2 infection, but the rRT-PCR shows a negative result and is not able to detect the viral genes. False-negative results in rRT-PCR are not uncommon (Hase et al. 2020; Kelly et al. 2020; Pan et al. 2020). A study on 51 patients of suspected viral pneumonia in Wuhan, only 71% were positive by rRT-PCR throat swab or sputum sample, while 98% had abnormal CT scan (Ai et al. 2020).

A negative COVID-19 test result may lead to a false sense of security. A SARS-CoV-2-infected person can transmit the virus and infect approximately three more persons (the reproductive number is averaged to be 3.28) (Liu et al. 2020). The Diamond Princess Cruise ship case study showed about 18% of cases to be asymptomatic (Mizumoto et al. 2020). Various other studies show that many cases are asymptomatic.

The SARS-CoV-2 is an enveloped virus with a positive-sense RNA genome encoding for about 27 proteins, including an RNA-dependent RNA polymerase (*RdRP*) and four structural proteins, including the spike surface glycoprotein (S), a small envelope protein (E), matrix protein (M), and nucleocapsid protein (N) (Udugama et al. 2020). The *RdRP* and *E* gene have conserved sequences and high analytical sensitivity for detection. So, these two genes are being used for detecting SARS-CoV-2 infection by rRT-PCR. There are various PCR kits and extraction kits available at present. Some claiming a sensitivity close to 100%. However, false-negative results can be frequent. Thermal inactivation of the sample, for example, can be a reason for false negativity (Pan et al. 2020). Mutations in the viral genes, including the *RdRP* gene, can be a cause of false-negative results in the RT-PCR (Pachetti et al. 2020; World Health Organization 2020b). In case of any discrepancy, following tests are being recommended by WHO.

2.1 Serological Tests

Serological tests detect the presence of antigen-specific antibodies present in the serum of the patient. These antibodies take about 1 week to develop, after the onset of infection and thus the test may be negative during the initial days of infection. A test detecting viral genes in the sample tells you if you are currently infected and is very sensitive if performed in the first week of acquiring the infection. An antibody test tells about the previous infection. Serological tests are more useful as epidemiological tools for detecting retrospectively the attack rate of the outbreak (World Health Organization 2020b).

2.2 Gene Sequencing

The World Health Organization recommends gene sequencing of some percentage of positive samples after NAAT for confirmation. This may help us to identify potential sites of mutations that might affect diagnostic tests.

3   Implications of a Negative COVID-19 test

As per the latest information (10th May 2020, 11:06 GMT), the highest number of COVID-19-positive cases (1,347,318 cases) have been detected in the United States of America after a total number of 8,918,345 tests (Worldometer 2020). It indicates that 84.89% of Americans, tested for COVID-19, are found negative on the test. As far as the COVID-19-negative test results are concerned, they are found to be variable in various countries (Table 1).

3.1   Diagnostic Implications

The World Health Organization recommends repeat testing especially in areas where the virus is widely spread (World Health Organization 2020b). If upper respiratory samples are showing a negative result in a patient with a high index of suspicion (World Health Organization 2020b), lower respiratory tract samples, like sputum and BAL, can be collected and tested. Induced sputum is not recommended (PIH 2020). In this scenario, other tests recommended by WHO, can also be considered.

Table 1 Countries with a percentage of negative COVID-19 tests

Country	Total number of tests done	Number of COVID-19-positive cases	COVID-19 negative percentage
Algeria	6500	5558	14.49%
Australia	809,616	6939	99.14%
Bangladesh	122,557	14,657	88.04%
Bolivia	7651	2437	68.15%
Brazil	339,552	156,061	54.04%
China	No information available	82,901	–
France	1,384,633	176,658	87.24%
Gabon	724	661	8.70%
India	1,609,037	62,939	96.09%
Iran	573,220	106,220	81.47%
Italy	2,514,234	218,268	91.32%
Nepal	72,239	110	99.85%
Nigeria	23,835	4151	82.58%
Pakistan	283,517	29,465	89.61%
Peru	494,250	65,015	86.84%
Russia	5,448,463	209,688	96.15%
South Africa	324,079	9240	97.15%
Spain	2,467,761	264,663	89.27%
United Kingdom	1,728,443	215,260	87.55%
United States of America	8,918,345	1,347,318	84.89%

As shown in Fig. 1, if a patient clinically suspected is tested negative then the fresh sample should be collected. The test may be reperformed using a different kit, if possible viral gene sequencing from the original sample of an amplicon generated from an appropriate NAAT assay. Laboratories should seek confirmation of the reference laboratory in case of any doubt (World Health Organization 2020b). If an asymptomatic person is tested negative for COVID-19 in rPCR, he could be true negative or could be an asymptomatic carrier. In the presence of travel history or history of contact (with a positive case), these people should be kept in isolation for 14 days and repeat samples to be collected on the fifth day.

3.2 Clinical Implications

An individual, who was found to be negative on COVID-19 testing, may have several clinical implications (Fig. 2).

An individual, who tests negative for COVID-19, can be symptomatic or asymptomatic. Symptomatic individuals can still have COVID-19 infection, despite the test being negative. Here, the fallacy lies with the sensitivity of the test. It is important to choose a highly sensitive test so that the false-negative cases will be less. A false-negative case is a potential threat to community transmission of infection. The same thing applies to asymptomatic individuals too. An asymptomatic false-negative individual is more dangerous. They can transmit infection as an asymptomatic carrier (Bai et al. 2020). Because of the negative test result, the perception of the individual as well as the society is expected to be more liberal. Asymptomatic individuals also have a higher chance of coming in contact with others in comparison to symptomatic individuals. If an individual is truly negative for COVID-19, that individual is currently not having the COVID-19 infection. If a true negative for COVID-19 is symptomatic (respiratory and other constitutional symptoms), then that individual needs to be evaluated for other medical conditions, that possibly attributing to such symptoms. During this pandemic, all patients with respiratory symptoms are predominantly evaluated in the line of COVID-19. The possibility of such evaluation increases, if there is a recent travel history or contact history. It is important to not miss a case of COVID-19 during the evaluation; at the same time, it is also important to consider the common differential diagnoses (alternative medical conditions). If the individual is asymptomatic and truly negative, then it is quite possible that the individual is healthy or having a medical condition, that is well controlled on treatment or lifestyle modifications. Several peculiarities with negative test results of COVID-19 have been observed. There have been reports of COVID-19 patients with negative test findings from the nasopharyngeal and sputum sample (Chen et al. 2020; Winichakoon et al. 2020; Zhang et al. 2020). One of these patients had a positive test result from the fecal sample, though the nasopharyngeal sample was negative (Chen et al. 2020).

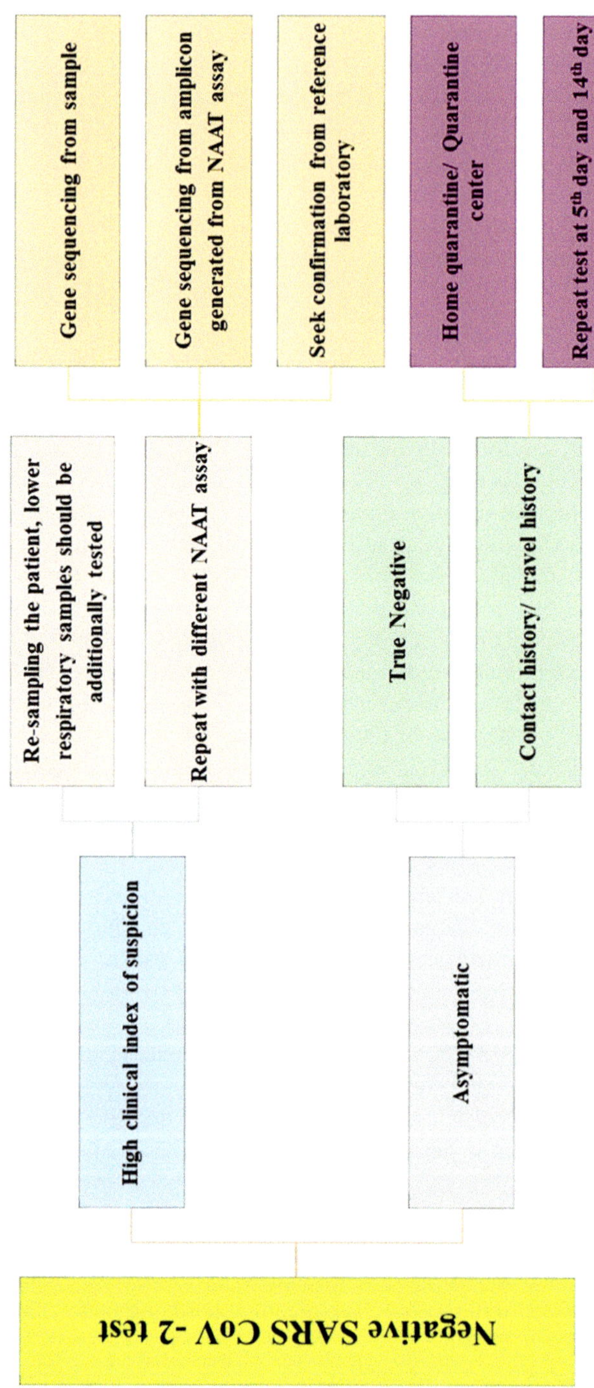

Fig. 1 Diagnostic implications of negative SARS-CoV-2 test

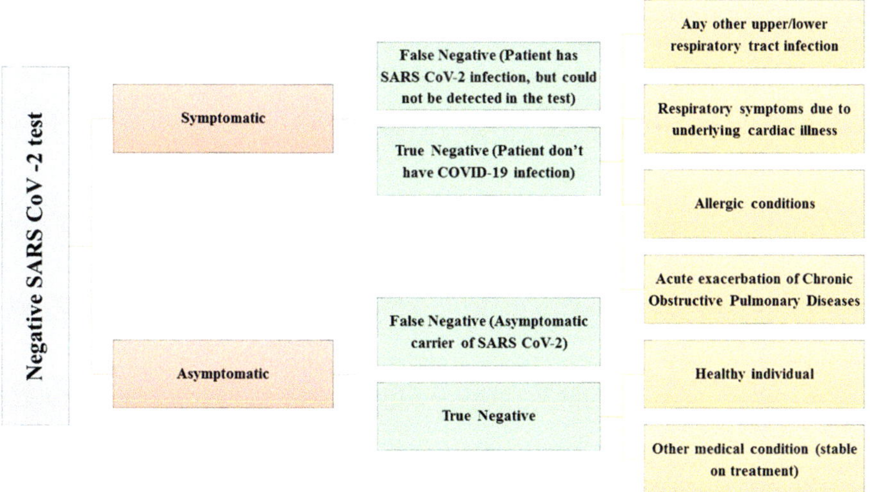

Fig. 2 Clinical implications of negative COVID-19 test

3.3 Mental Health Implications

In the global emergency of COVID-19, the importance of mental health has been emphasized strongly (Holmes et al. 2020). COVID-19 is affecting the mental health irrespective of gender, stage, and situation of life. It is affecting the psychological aspects the COVID-19 positive patients, their close contacts, friends, and families, as well as persons with physical or mental comorbidities, and the unaffected general population at large (Kar et al. 2020). However, the psychological reactions differ from individual to individual and situation to situation. The cases may develop wide range of abnormalities like fear, anxiety, depressive reactions/disorders, stress reaction/disorders, sleep disturbances, substance abuse, self-harm, and suicides (Holmes et al. 2020; Kar et al. 2020; Van Bavel et al. 2020; Wang et al. 2020). However, all the cited studies have evaluated the responses in a short period, as the pandemic is still in the evolving phase. Longitudinal, cross-national large scales studies are warranted to measure the psychological aspects in such patients. Along with the COVID-19-positive cases, the test-negative persons could have immediate as well as enduring effects on their mental health, which is yet to be revealed empirically. The psychological reactions are shaped by age, sex, education, social status, personality trait, religious belief and practice, economic capability, stress-coping style, experience, etc. Any pre-existing physical disease or mental disorder can intensify the reactions and increases the chance of death.

3.3.1 Reactions of Test Negative Persons Among the Healthy Population

This section deals with the reactions of people who get negative result in diagnostic tests. They can have reactions of denial regarding the test result, anxiety, safety-seeking behavior, repeated assurance seeking, and sleep disturbances. The psychological reaction varies depending on whether the person has active symptoms or not. Persons with nonspecific symptoms could have more propensity to develop more psychological symptoms. Furthermore, persons with anxiety traits and/or disorders, obsessive–compulsive disorder (OCD), a hypochondriac trait, may develop anxiety, panic symptoms and symptoms of OCD. Patients with pre-existing depressive disorder may develop anxiety, negative thoughts, pessimism, and nihilist thoughts. Test negative asymptomatic people usually feel secure and relaxed, whereas some people in this group may develop psychological symptoms due to some circumstantial factors, and personality patterns. The reactions of test-negative persons among the close contacts, family, and friends could make them delighted, relaxed, and secured, as they are not supposed to be isolated or quarantined (Wang et al. 2020). They may also become anxious, suspicious, and dubious regarding the test result. A negative test of healthcare providers could make them happy, as they have adequate knowledge regarding the test as well as the disease process. However, persons with anxious traits may develop anxiety symptoms.

3.3.2   Reactions of Test-Negative Persons Among the COVID-19 Cases

Becoming test negative is the acceptable outcome of the COVID-19 patients. As an expected event, the moment an individual is declared negative, the emotional reaction can be a sense of relief and happiness. However, it can also create anxiety regarding being re-infected and chances of getting the false-negative results.

3.3.3   Coping with Mental Health Issues After Becoming COVID-19 Test Negative

The coping strategies should be thought of for the psychological issues after a COVID-19 patient is declared test negative. The coping strategies can be either problem focused or emotional focused or both. Enhancing the coping measures is an important component of mental health promotion, which is expected to be beneficial during this pandemic. Mental health promotion activities mentioned as universal, selective, and indicated prevention should be established and implemented (Kar et al. 2020). Mass media can play a vital role. The government and public health authorities can disseminate necessary information (Kar et al. 2020; Kaufman et al. 2020). A specially arranged group therapy targeted to the negative-test individuals could be an important option (Van Bavel et al. 2020). An adequacy of mental health support systems and services should be ensured for the selected individuals (Jung and Jun 2020). Tele-mental health support services and specially designed smartphone apps can provide scientific information and reduce fear as well as improve mental well-being (Holmes et al. 2020).

3.4   Social Implications

Like any other individual, the COVID-19 test-negative individuals are also living in the society. Response of society is an important domain of well-being, and it affects the quality of life of these individuals. Varieties of response could be assumed from the community, which ranges from good acceptance to rejection in all the spheres of social life. It depends mainly on the level of knowledge of people regarding the virus, its transmission, the testability, and the credibility of the tests. However, individuals' social position, education, and personal influences may also affect it. Stigma regarding the diseases, controls the major part of the social reactions.

Adequate, appropriate, and culturally sensitive strategies should be targeted and implemented to reduce the social burden, where mass media, small group discussions could be the most effective decider. Community participation reduces the emotions of adverse reactions. The reduction of stigma should the prime target to control social adversities (Kar et al. 2020).

4   Conclusion

The COVID-19 test finding (positive or negative) is a dynamic phenomenon. The status conversion is a bidirectional process. An individual, who is positive or negative at a particular point of time, may have a different status at another point in time. A negative test result for COVID-19 is more common than positive result. Also, a negative test result is always not indicative of the absence of infection. A negative test result, in a symptomatic individual, opens the window for a range of possibilities.

5   Future Perspectives

As the pandemic is intensifying globally, increased number of diagnostic tests for COVID-19 is being done in most of the countries. There are clear guidelines to address the health issues of positive tested individuals; however, the guidelines are not clear about individuals, who are found to be negative. Though the guidelines recommend the subsequent testing protocol for individuals, who are tested negative, but the other healthcare needs also need to be incorporated into the guidelines. The guidelines are changing from time to time keeping in pace with the progress and outcomes during this pandemic. Future guidelines may give insight regarding approach to the individuals tested negative for COVID-19. Point-of-care tests like simple rapid test kits can detect proteins (antigens) from the COVID-19 patients in respiratory samples or can detect antigen-specific antibodies generated in response to infection in blood or serum (World Health Organization 2020c). Though these tests are not yet recommended by the WHO, research is underway. Researchers are also trying to develop biosensors with high specificity to detect SARS-CoV-2 in public places (Balfour 2020). Nanoparticles are also being tested as important detector molecules in place of fluorescence being used presently in rRT-PCTR, to increase sensitivity (Yee 2020).

6 Executive Summary

- Negative COVID-19 tests vary across countries, globally, and have a broad range of variation.
- True-negative test excludes current COVID-19 infection, whereas a false-negative test reflects the failure of the test to detect the COVID-19 infection.
- A highly sensitive test reduces the chances of getting a false-negative result.
- Negative COVID-19 test can have diagnostic, clinical, as well as psycho-social implications.

References

Ai T, Yang Z, Hou H, Zhan C, Chen C, Lv W, Tao Q, Sun Z, Xia L (2020) Correlation of chest CT and RT-PCR testing in coronavirus disease 2019 (COVID-19) in China: a report of 1014 cases. Radiology 200642

Bai Y, Yao L, Wei T, Tian F, Jin D-Y, Chen L, Wang M (2020) Presumed Asymptomatic Carrier Transmission of COVID-19. JAMA 323(14):1406–1407. https://doi.org/10.1001/jama.2020.2565

Balfour H (2020) Developing a COVID-19 biosensor for faster diagnostics and environmental monitoring. https://www.europeanpharmaceuticalreview.com/news/117659/developing-a-covid-19-biosensor-for-faster-diagnostics-and-environmental-monitoring/

Chen L, Lou J, Bai Y, Wang M (2020) COVID-19 Disease With Positive Fecal and Negative Pharyngeal and Sputum Viral Tests. Am J Gastroenterol 115. https://doi.org/10.14309/ajg.0000000000000610

Government of India (2020) Revised guidelines on clinical management of COVID – 19 [Ministry of Health & Family Welfare; Govt. of India]. https://www.mohfw.gov.in/pdf/RevisedNationalClinicalManagementGuidelineforCOVID1931032020.pdf

Hase R, Kurita T, Muranaka E, Sasazawa H, Mito H, Yano Y (2020) A case of imported COVID-19 diagnosed by PCR-positive lower respiratory specimen but with PCR-negative throat swabs. Infect Dis (London, England):1–4. https://doi.org/10.1080/23744235.2020.1744711

Holmes EA, O'Connor RC, Perry VH, Tracey I, Wessely S, Arseneault L, Ballard C, Christensen H, Cohen Silver R, Everall I, Ford T, John A, Kabir T, King K, Madan I, Michie S, Przybylski AK, Shafran R, Sweeney A et al (2020) Multidisciplinary research priorities for the COVID-19 pandemic: A call for action for mental health science. Lancet Psychiatry. https://doi.org/10.1016/S2215-0366(20)30168-1

Jung SJ, Jun JY (2020) Mental health and psychological intervention amid COVID-19 Outbreak: Perspectives from South Korea. Yonsei Med J 61(4):271–272

Kar SK, Yasir Arafat SM, Kabir R, Sharma P, Saxena SK (2020) Coping with mental health challenges during COVID-19. Coronavirus Dis 2019 (COVID-19):199–213. https://doi.org/10.1007/978-981-15-4814-7_16

Kaufman KR, Petkova E, Bhui KS, Schulze TG (2020) A global needs assessment in times of a global crisis: world psychiatry response to the COVID-19 pandemic. BJPsych Open:1–11

Kelly JC, Dombrowksi M, O'neil-Callahan M, Kernberg AS, Frolova AI, Stout MJ (2020) False-Negative COVID-19 Testing: Considerations in Obstetrical Care. Am J Obstet Gynecol Mfm. https://doi.org/10.1016/j.ajogmf.2020.100130

Liu Y, Gayle AA, Wilder-Smith A, Rocklöv J (2020) The reproductive number of COVID-19 is higher compared to SARS coronavirus. J Travel Med

Mahapatra S, Chandra P (2020) Clinically practiced and commercially viable nanobio engineered analytical methods for COVID-19 diagnosis. Biosens Bioelectron 165:112361

Mizumoto K, Kagaya K, Zarebski A, Chowell G (2020) Estimating the asymptomatic proportion of coronavirus disease 2019 (COVID-19) cases on board the Diamond Princess cruise ship, Yokohama, Japan, 2020. Eur Secur 25(10):2000180

Pachetti M, Marini B, Benedetti F, Giudici F, Mauro E, Storici P, Masciovecchio C, Angeletti S, Ciccozzi M, Gallo RC, Zella D, Ippodrino R (2020) Emerging SARS-CoV-2 mutation hot spots include a novel RNA-dependent-RNA polymerase variant. J Transl Med 18(1):1–9. https://doi.org/10.1186/s12967-020-02344-6

Pan Y, Long L, Zhang D, Yan T, Cui S, Yang P, Wang Q, Ren S (2020) Potential false-negative nucleic acid testing results for Severe Acute Respiratory Syndrome Coronavirus 2 from thermal inactivation of samples with low viral loads. Clin Chem. https://doi.org/10.1093/clinchem/hvaa091

PIH (2020) PIH Guide I COVID-19 Part I: Testing, contact tracing and community management of COVID-19. pp 1–23. Partners In Health. https://www.pih.org/sites/default/files/2020-04/PIH_Guide_COVID_Part_I_Testing_Tracing_Community_Managment_4_4.pdf

Udugama B, Kadhiresan P, Kozlowski HN, Malekjahani A, Osborne M, Li VYC, Chen H, Mubareka S, Gubbay JB, Chan WCW (2020) Diagnosing COVID-19: The disease and tools for detection. ACS Nano. https://doi.org/10.1021/acsnano.0c02624

Van Bavel JJ, Baicker K, Boggio PS, Capraro V, Cichocka A, Cikara M, Crockett MJ, Crum AJ, Douglas KM, Druckman JN (2020) Using social and behavioural science to support COVID-19 pandemic response. Nat Hum Behav:1–12

Wang C, Pan R, Wan X, Tan Y, Xu L, Ho CS, Ho RC (2020) Immediate psychological responses and associated factors during the initial stage of the 2019 coronavirus disease (COVID-19) epidemic among the general population in China. Int J Environ Res Public Health 17(5):1729

Winichakoon P, Chaiwarith R, Liwsrisakun C, Salee P, Goonna A, Limsukon A, Kaewpoowat Q (2020) Negative nasopharyngeal and oropharyngeal swabs do not rule out COVID-19. J Clin Microbiol 58:5. https://doi.org/10.1128/JCM.00297-20

World Health Organization (2020a) Coronavirus disease 2019 (COVID-19) situation report – 110. World Health Organization. https://www.who.int/docs/default-source/coronaviruse/situation-reports/20200509covid-19-sitrep-110.pdf?sfvrsn=3b92992c_4

World Health Organization (2020b) Laboratory testing for 2019 novel coronavirus (2019-nCoV) in suspected human cases. World Health Organization. https://www.who.int/publications-detail/laboratory-testing-for-2019-novel-coronavirus-in-suspected-human-cases-20200117

World Health Organization (2020c) Advice on the use of point-of-care immunodiagnostic tests for COVID-19. https://www.who.int/news-room/commentaries/detail/advice-on-the-use-of-point-of-care-immunodiagnostic-tests-for-covid-19

Worldometer (2020) COVID-19 Coronavirus pandemic. https://www.worldometers.info/coronavirus/

Yee KM (2020) Beyond RT-PCR: new methods emerge for coronavirus diagnostics. https://www.labpulse.com/index.aspx?sec=sup&sub=mic&pag=dis&ItemID=801078

Zhang P, Cai Z, Wu W, Peng L, Li Y, Chen C, Chen L, Li J, Cao M, Feng S, Jiang X, Yuan J, Liu Y, Yang L, Wang F (2020) The novel coronavirus (COVID-19) pneumonia with negative detection of viral ribonucleic acid from nasopharyngeal swabs: A case report. BMC Infect Dis 20(1):317. https://doi.org/10.1186/s12879-020-05045-z

Zhou P, Yang X-L, Wang X-G, Hu B, Zhang L, Zhang W, Si H-R, Zhu Y, Li B, Huang C-L (2020) A pneumonia outbreak associated with a new coronavirus of probable bat origin. Nature 579 (7798):270–273

Correction to: Overview of Coronavirus Disease and Imaging-Based Diagnostic Techniques

Archana Ramadoss, Veena Raj, and Mithun Kuniyil Ajith Singh

Correction to:
Chapter 5 in: P. Chandra, S. Roy (eds.), *Diagnostic Strategies for COVID-19 and other Coronaviruses*,
https://doi.org/10.1007/978-981-15-6006-4_5

The original version of this paper was published with errors. The below listed late errors have been corrected with this erratum.

1. Caption of Fig. 21: the incorrect citation has been replaced with the correct citation to read as:
   Fig. 21. Sonographic characteristics of moderate, severe, and critical pleural and parenchymal changes in COVID-19 patients. (Source: image reprinted from Smith et al. 2020)
2. The incorrect reference citations in the section 2.3, page 79, line 32-35 have been corrected to read as:
   The viral genome replication takes place in the cell cytoplasm at the periphery of the ER–Golgi intermediate compartment (ERGIC) (Balasuriya et al. 2017; Knoops et al. 2008; Ciulla 2020; Gosert et al. 2002).
3. The uncited references 72-75 have been deleted from the references list

The updated online version of this chapter can be found at
https://doi.org/10.1007/978-981-15-6006-4_5